A WORLD BANK COUNTRY STUDY

KOREA

Managing the Industrial Transition

Volume II
Selected Topics and Case Studies

The World Bank
Washington, D.C., U.S.A.

The World Bank
1818 H Street, N.W.
Washington, D.C. 20433, U.S.A.

Manufactured in the United States of America
First printing March 1987
Second printing December 1987

World Bank Country Studies are reports originally prepared for internal use as part of the continuing analysis by the Bank of the economic and related conditions of its developing member countries and of its dialogues with the governments. Some of the reports are published informally with the least possible delay for the use of governments and the academic, business and financial, and development communities. Thus, the typescript has not been prepared in accordance with the procedures appropriate to formal printed texts, and the World Bank accepts no responsibility for errors. The publication is supplied at a token charge to defray part of the cost of manufacture and distribution.

Any maps that accompany the text have been prepared solely for the convenience of readers. The designations and presentation of material in them do not imply the expression of any opinion whatsoever on the part of the World Bank, its affiliates, or its Board or member countries concerning the legal status of any country, territory, city, or area or of the authorities thereof or concerning the delimitation of its boundaries or its national affiliation.

The most recent World Bank publications are described in the catalog *New Publications*, a new edition of which is issued in the spring and fall of each year. The complete backlist of publications is shown in the annual *Index of Publications*, which contains an alphabetical title list and indexes of subjects, authors, and countries and regions; it is of value principally to libraries and institutional purchasers. The continuing research program is described in *The World Bank Research Program: Abstracts of Current Studies*, which is issued annually. The latest edition of each of these is available free of charge from the Publications Sales Unit, Department F, The World Bank, 1818 H Street, N.W., Washington, D.C. 20433, U.S.A., or from Publications, The World Bank, 66 avenue d'Iéna, 75116 Paris, France.

Library of Congress Cataloging-in-Publication Data

Korea : managing the industrial transition.

(A World Bank country study)
Report written by D.M. Leipziger and others.
1. Industry and state--Korea (South)
2. Korea (South)--Commercial policy. 3. Monetary policy--Korea (South) 4. Korea (South)--Economic policy. I. Leipziger, Danny M. II. International Bank for Reconstruction and Development. III. Series.
HD3616.K853K67 1987 338.9519'5 87-2213
ISBN 0-8213-0887-4 (v. 1)
ISBN 0-8213-0893-9 (v. 2)

PREFACE

This volume is composed of eight chapters dealing with various aspects of industrial policy in Korea. They provide further in-depth treatment of the issues of industrial structure, macroeconomic policy, trade and financial liberalization, and the economics of intervention, which affect the conduct of industrial policy in Korea.

Volume II contains five special-topic chapters. Chapter 1 addresses the issues of industrial transformation, using the latest (1983) input-output table as the basis of analysis, and also examines the issue of industrial concentration. Chapter 2 reviews the conduct of macroeconomic policy in the recent past, focusing in particular on structural changes which affect the relationship between instruments and targets. In Chapter 3, the issue of liberalization sequencing is addressed and the risks of the Latin American experience are reviewed. Chapter 4 includes both a conceptual treatment of the theory of industrial interventions and applications for Korea as well as a summary of recent Japanese restructuring activities. Chapter 5 examines the link between industrial performance and access to financing.

This volume also contains three industrial case studies. Chapter 6 examines the shipping industry and focuses on how government has intervened to manage its restructuring. Chapter 7 deals with the important textile industry, which is undergoing considerable transformation. Chapter 8 addresses the emerging industry case of electronics, on which Korea planners are placing a good deal of faith as a future leading industry.

SPECIAL TOPICS AND CASE STUDIES

Table of Contents

This report was prepared on the basis of missions visiting Seoul in August and December, 1985 and was written by D.M. Leipziger (Mission chief), Y.J. Cho, F. Iqbal, P. Petri, and S. Urata with contributions by S. Edwards (Consultant) and D. O'Connor and the assistance of S.Y. Song. Administrative assistance in Washington was provided by J. Tolbert.

Page No.

Page No.

Page No.

List of Tables

Page No.

Page No.

Page No.

List of Figures

List of Boxes

CHAPTER 1: KOREA'S INDUSTRIAL STRUCTURE

A. Introduction

1.01 The purpose of this chapter is to briefly review the major changes in industrial structure that have occurred since 1970. Part I examines industrial structure per se, utilizing the latest (1983) input-output relationships and analyzes how the economy, especially the manufacturing sector, has changed since 1970. Part II looks specifically at the import substitution phase and attempts to identify changes in Korea's industrial structure that emerged from this controversial policy, as well as the linkages between HCI, import dependency and export development. In Part III the future structure of Korean industry is reviewed and several policy issues, namely employment, productivity and R&D are identified. Part IV is devoted to the issue of industrial concentration, an issue of rising importance in the Korean context, particularly insofar as conglomerates are concerned. Finally, Appendix 6A examines the Korean experience in the context of the international comparisons of growth framework, while Appendix 1B looks at Korea's historical sources of growth from both the supply and demand perspectives.

B. Structural Changes in the Economy

Shifts in Production

1.02 The structure of production continued shifting from agriculture to manufacturing throughout the 1970-83 period, regardless of the indicator (output, value added, or employment) of measurement. According to output structure, for example, the share of agriculture declined drastically from 17% in 1970 to 8.3% in 1980, but remained virtually unchanged since then (Table 1.1). On the contrary, the share of manufacturing increased sharply from 40.3% in 1970 to 50.4% in 1975, and maintained that level through 1983. Despite the similarity in the direction of change in production structure measures, the variation in the size of sectoral shares is quite different depending on the indicator chosen. In 1983 agriculture contributed 29.3% of employment and 14.1% of value added, while its share of output was only 8.2%. In contrast, manufacturing captured 50.0% of output in 1983, while it only accounted for 22.2% of employment and 29.5% of value added.

Trends in Manufacturing

1.03 Among the manufacturing subsectors, light industry, after gaining marginally between 1970 and 1975, saw its share of total output fall considerably between 1975 and 1983. Heavy industry, on the other hand, saw its share of gross output almost double between 1970 and 1975 and rise significantly again by 1980 (Table 1.2). According to the international comparison of industrial development, reported in Appendix 1A, the share of heavy industry started to increase rapidly in 1973-74, exceeding normal predicted levels. As a result of these contrasting developments, heavy industry surpassed light industry in its share of total output by 1980. Although the shares of all three subsectors (chemicals, primary metals and machinery) in HCI increased between 1970 and 1983, there are some differences in their developments between pre- and post-1980 periods. After recording a substantial gain between 1970 and 1980, chemical and chemical products and primary metal manufacturing lost ground slightly between 1980 and 1983, while metal products and machinery continued expanding their shares even after 1980.

Table 1.1: STRUCTURAL CHANGE IN PRODUCTION
(Percentage shares)

	Gross output				Value added				Employment			
	1970	1975	1980	1983	1970	1975	1980	1983	1970	1975	1980	1983
Agriculture	17.0	12.8	8.3	8.2	25.2	22.7	14.7	14.1	50.2	41.4	32.0	29.3
Mining	1.1	0.9	0.8	0.7	1.7	1.5	1.4	1.1	1.1	1.1	1.1	1.1
Manufacturing	40.3	50.4	51.0	50.0	20.1	26.1	28.2	29.5	12.4	19.2	21.7	22.2
Construction	8.6	6.2	8.0	8.2	6.6	5.2	8.3	8.4	3.7	4.0	5.3	5.6
Social overhead	6.7	6.7	8.1	8.9	8.7	7.5	9.9	11.4	3.8	4.3	4.5	4.9
Services	26.3	23.0	23.8	23.9	37.7	36.9	37.5	35.4	28.7	30.0	35.3	37.0

Source: Bank of Korea, Input-Output Tables.

1.04 A detailed examination of the developments in the metal products and machinery subsector during 1980-83 shows that the increase in the composition share was particularly substantial in general machinery, electrical machinery and transportaion equipment (see Table 1.3). These industries have also provided a significant number of new jobs since 1980 (data) and are expected to be among the fastest growing industries for future labor absorption.[1]

1.05 Unlike the developments in heavy industry subsectors, the direction of the composition shares among the light industry subsectors was less uniform. Within light industry, food, beverage and tobacco as well as lumber and wood saw their output share decline continuously throughout the period. Textiles and leather did well in the first half of the 1970s but its share has declined considerably since then, to a point in 1983 where its share of gross output was essentially on par with its 1970 share. The industry continues to

---

1/ According to KDI estimates, it is expected that the electric and electronics industry will have grown from 10.8% of manufacturing to 14.6% by 1991 and 18.3% by 1996. Smaller but still significant increases are projected for machinery and transport machinery, so that these emerging sectors taken together are expected to increase their combined share of manufactures from 25.3% (1983) to 36.9% (1996).

Table 1.2: STRUCTURAL CHANGE IN MANUFACTURING
(Percentage shard in total output)

	Gross output				Value added				Employment			
	1970	1975	1980	1983	1970	1975	1980	1983	1970	1975	1980	1983
Light industry	28.4	29.5	24.7	22.1	12.8	14.5	13.7	13.6	9.2	13.5	13.8	13.0
Food, beverages and tobacco	15.9	14.4	10.8	9.6	6.2	6.3	6.2	5.9	2.5	2.8	2.9	3.0
Textiles and leather	7.1	9.9	8.4	7.0	3.8	5.5	4.9	3.9	4.0	7.9	7.4	6.5
Lumber and wood products	1.4	1.2	1.0	0.9	0.7	0.5	0.4	0.4	0.6	0.6	0.7	0.6
Paper printing and publishing	1.4	1.4	1.6	1.8	0.4	1.3	1.1	1.2	0.6	0.8	0.9	1.0
Nonmetallic metal manufacturing	1.4	1.5	1.9	1.8	1.1	1.3	1.4	1.4	0.6	0.7	0.9	0.9
Miscellaneous manufacturing	1.2	1.1	1.0	1.0	1.0	0.9	0.8	0.8	0.9	0.7	1.1	1.1
Heavy and chemical products	11.9	20.9	26.3	27.9	7.2	11.6	14.5	15.9	3.2	5.7	7.9	9.2
Chemical and chemical products	5.9	10.8	12.6	11.8	4.0	5.9	6.7	6.3	1.2	1.9	2.5	2.7
Primary metal manufacturing	2.0	3.4	5.1	5.0	0.7	1.0	1.7	1.8	0.4	0.5	0.7	0.9
Metal products and machinery	4.0	6.7	8.6	11.2	2.5	4.7	6.1	7.9	1.6	3.3	4.7	5.6

Source: Bank of Korea, Input-Output Tables

Table 1.3: SIGNIFICANT CHANGES IN THE METALS AND MACHINERY INDUSTRY
(percentage shares in total output)

	Gross output				Value added				Employment			
	1970	1975	1980	1983	1970	1975	1980	1983	1970	1975	1980	1983
Metal products and machinery	4.0	6.7	8.6	11.2	2.5	4.7	6.1	7.9	1.6	3.3	4.7	6.3
Fabricated metal	-	(0.9)	(1.3)	(1.7)	-	(0.5)	(0.8)	(1.0)	(0.4)	(0.5)	(0.7)	(1.2)
General machinery	-	(0.8)	(1.4)	(2.1)	-	(0.6)	(1.2)	(1.5)	(0.3)	(0.5)	(0.8)	(1.0)
Electrical equipment	(2.2)	(1.0)	(1.2)	(1.4)	(1.3)	(0.7)	(0.9)	(1.0)	(0.2)	(0.4)	(0.6)	(0.7)
Electronic and communication equipment		(2.0)	(2.5)	(2.9)	-	(1.4)	(1.7)	(1.9)	(0.2)	(0.9)	(1.5)	(0.7)
Transportation equipment	(1.7)	(1.9)	(1.8)	(2.7)	(1.1)	(1.3)	(1.2)	(2.2)	(0.5)	(0.9)	(0.8)	(1.4)
Measuring, medical equipment	(0.1)	(0.3)	(0.4)	(0.3)	(0.08)	(0.1)	(0.3)	(0.2)	(0.07)	(0.1)	(0.3)	(0.3)

Source: Bank of Korea, Input-Output Tables.

be important as a source of employment, however, providing the largest share of manufacturing employment, followed by metals and machinery, which has exhibited the fastest growth in employment since 1970 among subsectors. Light industry's share is expected by Korean planners to continue to decline, in large measures because the share of manufacturing stemming from textiles is projected to decline from 11.0% in 1983 to 8.3% in 1996. This has significant employment implications, as noted in the employment section.

Increasing Use of Intermediate Inputs

1.06 Industrialization is usually accompanied by an increase in the use of intermediate inputs in production as measured by the intermediate inputs share. This results from the increasing specialization of economic activity, especially the enlarged share of manufacturing, which uses intermediate goods intensively. The intermediate inputs share in Korea increased continuously from 47.9% in 1960 to 50.4% in 1970, and then to 60.4% in 1980, but it declined slightly to 59.5% in 1983. The rate of increase accelerated in 1970-80, and it was especially rapid between 1970 and 1975, reflecting rapid heavy-chemical industrialization.[2/] A comparative perspective is given in Figure 1.1. Although both Korea and its major competitor experienced a parallel increase in the intermediate inputs share in the 1965-70 period, reflecting the fact that both countries were in the process of rapid industrialization, the rate of increase in Korea between 1970 and 1980 was greater, reflecting its capital-intensive strategy. In contrast to these developments, the intermediate inputs share in Japan declined continuously, probably indicating that Japan was already in the stage where the service sector had started expanding at the cost of manufacturing.

1.07 The increase in the intermediate inputs share indicates deepening of interindustry linkages, an important characteristic of the development process.[3/] Historical interindustry linkages in Korea have been much higher than in other developing countries with a similar level of per capita GNP, such as Turkey and Mexico, and much closer to the pattern observed in more developed countries such as Japan (see Table 1.4).[4/] A closer look at the characteristics of interindustry linkages over the 1963-1973 period reveals an important difference between those in Korea and its major competitor on the one hand and those in Japan on the other, however. Although overall linkages in these three countries are comparable, the size of domestic linkages in

---

2/ The increase in the intermediate inputs share was not affected by oil price increases. The rate of increase in the share of intermediate inputs share in constant prices (51.5% in 1970 and 60% in 1983) is similar to that in current prices.

3/ See, for example, Chenery, Robinson, and Syrquin, Industrialization and Growth: Comparative Study, forthcoming.

4/ Interindustry linkages here are computed as the amount of intermediate inputs (direct and indirect) generated in the process of production in order to meet 100 units of final demand.

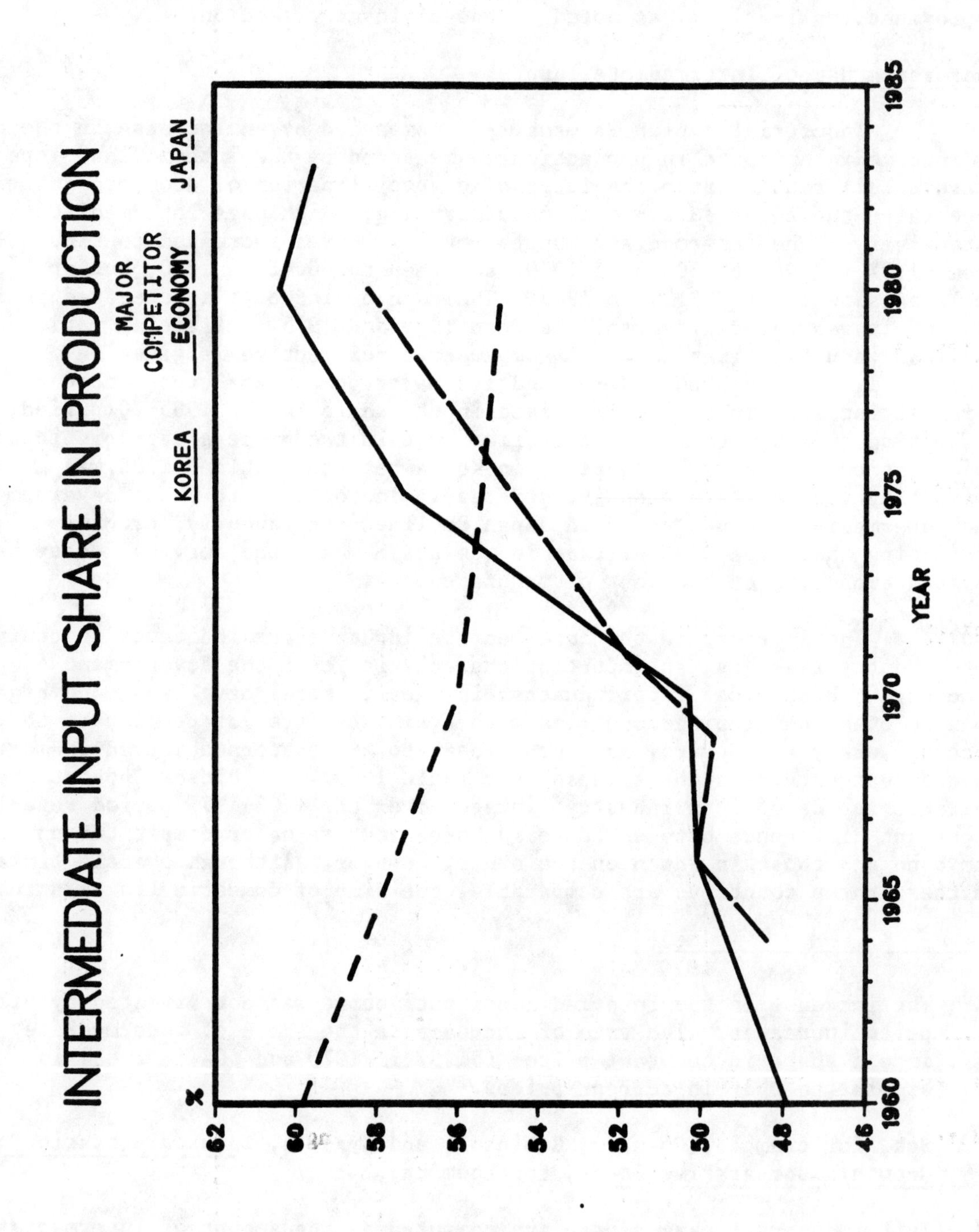
INTERMEDIATE INPUT SHARE IN PRODUCTION
KOREA
MAJOR COMPETITOR ECONOMY
JAPAN
%
62
60
58
56
54
52
50
48
46
1960
1965
1970
1975
1980
1985
YEAR

Figure 1.1

Table 1.4: INTERINDUSTRY LINKAGES
(Units of production generated to meet 100 units of final demand)

		Overall linkages	Domestic linkages
Korea	1963	89.9	60.9
	1970	89.8	58.7
	1973	92.8	54.6
Korea /a	1970	97.5	64.9
	1975	119.2	72.6
	1980	134.4	79.2
	1983	133.0	80.3
Major Competitor	1956	76.5	42.6
	1961	85.9	55.0
	1966	92.9	55.7
	1971	93.7	55.2
Turkey	1963	52.1	46.4
	1968	56.7	51.5
	1973	59.6	52.8
Mexico	1950	54.3	40.5
	1960	68.9	51.3
	1970	63.9	52.0
	1975	69.5	54.2
Japan	1955	89.9	81.3
	1960	94.5	82.7
	1965	74.6	82.4
	1970	106.3	88.7

/a Bank staff estimates based on Korean input-output data.

Source: Chenery, Robinson and Syrquin (forthcoming).

Table 1.5: TOP TEN EXPORTS

	1970	1975	1981	1985
1.	Textiles and Garments	Textiles and Garments	Textiles and Garments	Textiles and Garments
2.	Plywood	Electronic Products	Electronic Products	Ships
3.	Wigs	Steel Products	Steel Products	Electronic Products
4.	Minerals	Plywood	Footwear	Steel Products
5.	Electronic Products	Footwear	Ships	Footwear
6.	Fruits and Vegetables	Deep-sea Fish	Machinery	Machinery
7.	Footwear	Ships	Synth. Resin Products	Fish and Fish products
8.	Tobacco	Metal Products	Auto Tires	Synth. Resin Products
9.	Steel Products	Petroleum Products	Metal Products	Automobiles
10.	Metal Products	Synth. Resin Products	Plywood	Electric Products

Sources: Korea Traders Association, The Trend of Foreign Trade; Bank of Korea, Monthly Statistical Bulletin.

Korea, for example, which measures interindustry relationships of only domestic goods exclusively, is only about 70% of that in Japan. Furthermore, domestic linkages in Korea are not much different from those in Turkey or Mexico, although Korea's is the only declining trend. These findings indicate that Korea achieved a high level of interindustry linkages (implicitly greater technology interaction) initially through importation of intermediate goods. Using input-output data, an extension of this approach for Korea shows a significant increase in domestic industrial linkages between 1970 and 1980, although these interactions are still smaller than the newly generated international linkages over the period.

Changes in Trade Structure

1.08 Changes in Korea's export profile can be seen from a variety of perspectives. A very crude glimpse can be obtained by comparing Korea's top ten export items over time (see Table 1.5). Whereas textiles and garments are still the largest source of foreign exchange, some of the light industrial manufactures, like wigs and plywood, are now moribund, compared with 1970, having been replaced by capital-intensive products like ships and steel. Looking more broadly, the capital-labor ratio for manufactured exports increased between 1970 and 1983 from 1.6 to 6.3 million won per employee in 1977 prices, while the corresponding ratios for import substitutes were 2.1 and 7.5 million won (Table 1.6).[5/] Although import substitutes were more capital intensive than exports even in 1983, the gap between them narrowed between 1970 and 1983. A sources decomposition of the changes in capital-labor ratios in exports reveal that about 80% of the increase between 1970 and 1983 is due to the increase in overall capital-labor ratios, and the remaining 20% is due to the shift in the pattern of exports toward more capital intensive products. The findings here seem to indicate that the rapid increase in the capital-labor ratios in the Korean manufacturing exports is attributable not only to rapid capital formation due to active industrialization policies but also to incentives given to capital intensive exports.

Table 1.6: CHANGES IN CAPITAL-LABOR RATIOS
(in million won in 1977 prices)

	1970	1975	1980	1983
Domestic output (D)	2.21	3.09	5.76	7.49
Exports (E)	1.61	2.54	5.05	6.32
Import substitutes (MS)	2.10	3.38	5.88	7.46
E/MS	0.77	0.75	0.86	0.85

Source: World Bank estimates.

5/ Wontack Hong (1985) finds similar results by computing capital-labor ratios incorporating both direct and indirect factor requirements. In Japan in 1965, production of exports turned more capital intensive compared to the production of import substitutes in Japan.

1.09 Comparing the export patterns of Korea and Japan in the 1960s and 1970s, Korea can be seen to be exporting commodities which did not require a high level of human or physical capital intensity while Japan exported relatively human and physical capital intensive products.[6/] Aspects of Korea's export structure have been changing rapidly over time, however, as Korean exports have become more physical and human capital intensive very rapidly. As seen in Table 1.7, for example, which reports changes in "revealed comparative advantage," Korea's exports have been experiencing relative declines in traditional industries, such as textiles and cottons, wood products and paper, and basic assembly products. That is not to say that light manufactured exports are generally uncompetitive, as seen by the continued resiliency of clothing, footwear, luggage and cutlery exports. At the same time, Korean comparative advantage has clearly emerged over the last 10 years in shipbuilding, electrical equipment, metal products, and iron and steel--all extentions of the HCI sector. Although costly to establish, these capital-intensive industries have now reached the level of international competitiveness, and may legitimately be described as successful infant industries. On the other hand, in some industries, prominantly in the chemicals subsector and in machinery, Korea has yet to establish itself in international markets.

1.10 Korea appears to have made strong progress in shifting its export structure from labor-intensive to capital-intensive and moderately skill-intensive products. An interesting example of the underlying industrial flexibility at the firm level is seen in Box 1.1. More broadly, the initial transition to capital-intensive exports owes its success and speed to the interventions of the 1970s in establishing industries of sufficient scale to be internationally competitive. The 5 fastest growing industries between 1977 and 1981 in Korea's export profile, tellingly, were steel rods, containers (for shipping transport), household electrical equipment, tubes and pipes, and autos. These are industries in which scale and basic technology are important. The next step, to higher-technology intensive and skill-intensive products will be another milestone for Korean exports.

1.11 There is some reason to believe that this second transition will take longer to achieve. Why might this be the case? First, Government, chastened by the difficulties the economy experienced as a result of the HCI experience, is perhaps less likely to attempt another aggressive strategy in establishing skill and technology intensive industries. Second, accumulation of human capital takes longer than physical capital. Third, assuming that the role of the government in investment allocation is diminished in keeping with the financial liberalization goals discussed in Chapter 5 of Volume I, technology investment may not be expected to increase at the rate similar to that achieved for physical investment.

6/ Y.S. Lee, (forthcoming).

Table 1.7: KOREA'S REVEALED COMPARATIVE ADVANTAGE /a

Selected manufacturing items (SITC)	1975	1980	1984
Declining Advantage			
Textile yarn and thread (651)	1.159	0.996	0.648
Cotton fabrics (652)	0.462	0.219	0.139
Special textiles (655)	0.584	0.483	-0.053
Wood manufacturers (632)	0.462	-0.177	-0.554
Printed matter (892)	-0.776	-2.414	-2.570
Sound recorders (891)	0.838	0.103	-0.457
Office supplies (895)	0.876	0.812	-1.467
Comparative Disadvantage			
Paperboard (641)	-1.906	-1.339	-2.654
Cosmetics (553)	-3.768	-5.032	-3.798
Paints (533)	-3.311	-2.655	-2.155
Agricultural machinery (712)	-4.573	-3.811	-4.061
Road vehicles (732)	-5.109	-2.907	-2.775
Metalwork machinery (715)	-4.213	-1.934	-2.333
Continuing Advantage			
Clothing (841)	2.103	1.604	1.392
Fur and leather (842)	1.694	1.495	1.384
Footwear (851)	1.514	1.526	1.535
Luggage (831)	2.236	1.738	1.891
Cutlery (696)	1.605	1.439	1.302
Emerging Advantage			
Ships and boats	-0.096	0.916	1.808
Electrical equipment (725)	-2.481	-0.807	0.196
Metal products (691)	-2.278	0.055	1.044
Iron and steel	-0.563	0.316	0.191

/a RCAs are calculated here in natural logarithms. Therefore, a positive RCA implies that Korea's percentage share of a commodity export item compared to world exports of that item exceeds Korean manufacturing's total share in world exports.

Source: M. Noland (1985).

Box 1.1: KOREA'S INDUSTRIAL FLEXIBILITY AT THE FIRM LEVEL

In the 1971 the Handok Company was a wig manufacturer, with this one export item accounting for 95% of total sales. By 1976, Handok had diversified extensively, to the point where wigs were only 16% of sales and paper products made up 51% of output, complemented by tuna (22%) and the emergence of a new item, watches (9%). The industrial transformation was completed by 1981, when watches accounted for 85% of sales. And by 1985, liquid crystal display manufacturing, including monitors and dashboard items for example, were beginning to emerge as new sales items (10% of sales) and the bulk of revenue was due to computers and electronics (41%) and watches (45%). Handok is an example of industrial flexibility in a medium-sized firm employing about 3,500 people and generating sales of about 64 billion won in 1984.

C. Import Substitution in the HCI Sector

Reducing Import Dependency

1.12 It is widely known that the government pursued heavy-chemical industry (HCI) promotion policies through various means. Fiscal and financial policies were generously applied in the mid to late 1970s, while import protection policies in the form of quantitative restrictions and tariff protection, were also increased during the period. This section examines how effective the HCI promotion policies were in reducing the import dependency of the heavy and chemical industries. Table 1.8 indicates that import substitution took place at a rapid rate throughtout 1970-83 in virtually all the subsectors in the heavy-chemical industry as shown by a substantial decline in the import dependency of demand, defined as the share of imports in total demand (i.e., intermediate and final demand). In particular, the rate of import substitution was remarkably high in the primary metal manufacturing, and metal products and machinery industries.

1.13 Turning to import dependency in production, defined as the share of imported intermediate inputs in total inputs including value added, one finds that the reliance on imported intermediate inputs in the production of heavy-chemical products increased between 1970 and 1975, but has been declining slowly and continuously since then. The direction and the magnitude of the changes in the degree of import dependency was not uniform among the subsectors, however. In primary metal manufacturing the import dependency declined substantially throughout the period, while in metal products and machinery it increased slightly between 1970 and 1975 before beginning a steady decline from 1975 to 1983. In contrast, the import dependency in chemicals increased continuously throughout 1970-83 probably attributable to the rising price of oil. The decline in import dependency in primary metal and metal products and machinery is directly attributable to import substitution in these industries inasmuch as these industries acquired the capacity to produce more of their own inputs.

Table 1.8: IMPORT DEPENDENCY IN THE HEAVY-CHEMICAL INDUSTRY /a

	Demand				Production			
	1970	1975	1980	1983	1970	1975	1980	1983
Heavy and chemical industry	37.1	29.5	23.7	22.1	29.1	33.7	31.1	29.1
Chemical and chemical products	23.8	19.6	15.0	16.5	30.5	40.1	42.2	44.0
Primary metal manufacturing	35.3	27.6	18.9	17.0	36.0	30.1	20.2	17.6
Metal products and machinery	50.6	41.7	35.8	29.1	23.6	25.3	21.1	18.5
Light industry	8.0	8.1	7.2	7.7	12.3	13.8	13.7	13.4
All industry	11.3	14.6	14.8	13.9	8.6	12.8	14.2	13.5

/a Import dependency in demand is computed as inputs/total demand (final and intermediate demand), and import dependency in production as imported intermediate inputs/ total imports including value added.

Source: Bank of Korea, Input-Output Tables.

1.14 So far the degree of dependency has been examined by considering only direct requirements of imported intermediate in the production. Import dependency can also be measured by considering indirect requirements which would arise from interindustry linkages. Table 1.9 shows the amount of direct and indirect imports required to produce one unit of output. The import requirement for heavy and chemical industry declined slightly in 1980-83, but compared to light industry the production of HCI products requires a significantly higher amount (about 15% in 1983) of imports. Despite the fact that the import dependency in HCI has been declining rapidly, its reliance on imported inputs is still significantly higher than that of the heavy-chemical industry in Japan.[7/] One might expect the import dependency to continue to decline in the future as HCI continues to gain in efficiency, but such a development will have to continue in the context of import liberalization and thus be subject to the yardstick of international efficiency. According to the import liberalization schedule, almost total elimination of restrictions will occur for metal products and iron and steel products by 1987 and for electronics, electrical and non-electrical machines by 1988. This means that continued import substitution will have to be efficient by international standards.

7/ The import share in heavy-chemical industry in Japan was 4.3% in 1983. It is not only heavy-chemical industry that has high import dependency in the Korean economy. According to Chenery, Robinson and Syrquin (1986), Korean import requirements in production are much higher than other countries.

Table 1.9: DIRECT AND INDIRECT IMPORT REQUIREMENTS IN PRODUCTION (%)
(percent)

	1975	1980	1983
Heavy and chemical industry	48.7	49.0	45.8
Basic chemicals	47.4	53.6	48.4
Petroleum products	71.3	80.7	81.0
Primary iron and steel products	58.9	49.1	43.5
Fabricated metal products	44.4	40.1	36.7
General industrial machinery and equipment	35.0	35.1	35.3
Electronic and communication equipment	49.4	44.6	43.1
Transportation equipment	41.8	42.4	34.6
Light industry	26.4	30.0	28.7

Source: Bank of Korea, Quarterly Economic Review, December 1985.

Foreign Markets for Heavy-Chemical Products

1.15 The share of HCI output sold abroad (the export ratio) increased substantially from 7.4% in 1970 to 22.7% in 1983; a large increase of almost 10 percentage points took place between 1970 and 1975 (Table 1.10). All three subsectors (chemicals, primary metal and machinery) contributed to this rapid increase, but the magnitude of the contribution appears to have been especially high in metal products and machinery.[8] Despite the rapid growth in the early 1970s, the rate of increase in the export ratio slowed down in all three subsectors after 1975. In 1983, the export ratios in chemicals, primary metal and machinery were 12.1%, 20.1% and 35.1%, respectively. Relatively low export ratios in chemicals and primary metal manufacturing seem to be due to the fact that they are intermediate goods used domestically for

8/ Within metal products and machinery, transportation (mainly ships) appears to have contributed most to the rapid increase in the export ratio between 1970 and 1975.

the production of metal products and machinery, more than one-third of which is ultimately exported.[9] Among the machinery subsectors, the ratios in electronic and communications equipments and transportation equipments were especially high at 47.1% and 46.3%, respectively, in 1983. These ratios are slightly higher than the ratio observed in textiles and leather, and it is only slightly lower than the ratio for miscellaneous manufacturing such as toys, sporting goods and wigs.

Table 1.10: CHANGES IN EXPORT RATIOS
(percent)

	1970	1975	1980	1983
Heavy and chemical industry	7.4	17.2	19.3	22.7
Chemical and chemical products	6.2	10.5	10.4	12.1
Primary metal manufacturing	5.5	14.5	20.9	20.1
Metal products and machinery	10.0	29.1	31.6	35.1
Light industry	11.9	19.2	19.0	19.2
Textiles and leather	25.8	36.4	37.7	44.1
Miscellaneous manufacturing	54.1	58.6	53.4	53.6

Source: Bank of Korea, Input-Output Tables.

1.16 These results mirror those of the demand side sources of growth decomposition reported in Appendix 1B, which indicates that export expansion provided the most significant impetus to output growth not only for light industry but also for heavy and chemical industry in the 1970-83 period. Moreover, the contribution of export expansion to output growth in HCI accelerated over time--from 35.9% in 1970-75 to 41.3% in 1975-80, and then to 54.1% in 1980-83. Export incentives through financial and trade policies played an important role in the rapid expansion in the 1970s, while the depreciation of won provided an additional incentive in the 1980s. Observing a continuous increase in the magnitude of the export contribution to output growth leads to the query whether such a trend can continue in the future or whether domestic

---

9/ The indirect export ratios (outputs used as intermediate inputs in the production of exports) would be significantly higher than the direct export ratios for chemicals and primary metals. This can be seen by comparing the contribution of export expansion in the demand sources calculation between direct and total measures. In chemicals and primary metals the contribution computed by the total measure is significantly larger than that computed by the direct measure.

demand must at some point take on larger significance. The issue needs to be addressed in the context of the year 2000 projections.

Overview

1.17 This section has viewed the HCI episode from the trade perspective, namely, how was import dependency affected and how export-intensive was HCI production. The rationale is that Korea as a small, very open economy is unlikely to pursue an industrial strategy without an ultimate export aim. Thus, to merely characterize the HCI period as a classic import substitution experiment is incorrect. There is no doubt that the large-scale investments in HCI were distortionary of capital market (see Chapter 4 of Volume II) and economically costly in the medium-run, as seen by Table 1.11 on utilization.[10/] Yet, the policy was unique in other respects. First, while its aim was to establish Korean capacity in some basic industries, the investment program itself was fairly import-intensive.[11/] Second, the HCI sector did manage to export, indicative either of its ability to compete internationally or reflective of heavy subsidization on the capital side. Export ratios, seen in Table 1.10, show that all three subsectors had already experienced considerable export gains by 1975.[12/] Although further gains have been uneven, the sector as a whole now exports directly almost 23% of its production, more than the comparable figure for light industry. Moreover, while the direct import dependency of HCI production has fallen since 1975 (in subsectors other than chemicals), Korea's overall import requirements are still quite high, and attest to it being highly integrated with the global economy.

10/ The problem of underutilized capacity was especially acute in some subsectors, like fabricated metals and machinery, which were characterized by capacity utilization rates of 55-65% during 1976-79, considerably below the average in manufacturing as a whole.

11/ This helps to explain the finding in Appendix B on the sources of industrial growth that HCI growth appears to have been fueled primarily by export expansion and relatively modestly by import substitution factors.

12/ This finding is supportive of the evidence presented in Appendix A, which indicates on the basis of international norms that Korea's big surge in pushing HCI above the pattern expected occurred between 1970 and 1974.

Table 1.11: CAPACITY UTILIZATION BY INDUSTRIAL CLASSIFICATION (%)
(percent)

	1976	1977	1978	1979	1980	1981	1982	1983	1984	1985
Manufacturing	78.9	81.7	88.3	82.1	69.5	70.3	69.4	75.8	80.4	79.2
Light industry										
Food and beverages	62.4	75.8	85.6	82.4	69.3	64.3	64.9	75.4	73.8	73.0
Textiles and leather	87.5	85.9	84.8	82.1	80.1	80.9	80.2	79.0	78.9	76.7
Wood products	84.6	94.4	99.0	84.6	61.0	39.9	46.6	54.4	51.4	55.4
Paper products	72.8	80.8	88.4	85.1	75.4	74.8	72.5	76.4	82.1	79.5
Nonmetallic mineral products	81.9	88.2	87.3	77.9	63.6	61.1	68.3	77.6	78.4	72.6
Heavy industry										
Chemical products	91.9	98.1	110.4	95.4	80.3	76.0	70.9	75.4	78.5	79.6
Basic metal	78.6	81.1	88.1	81.0	71.3	71.2	74.7	83.8	87.3	88.9
Fabricated metal products, machinery	61.0	57.1	61.7	62.6	53.1	61.0	60.0	67.9	77.9	75.5

Note: Seasonally adjusted figures.

Source: Ministry of Trade and Industry

D. Future Structural Issues

Future Projections

1.18 Before closing the section on the patterns of change in economic structure, let us examine briefly how the Korean government projects its production structure to evolve in the future. Projections are examined at two different levels: one covering the entire economy, the other covering only the manufacturing sector.[13/] Table 1.12 reports KDI projections for production structure of GDP in years 1990 and 2000, as well as the corresponding Japanese shares in 1966 and 1982.[14/] The projected pattern reveals a decline in agriculture's share and offsetting increases in the other three sectors. Despite the continuing increase in the manufacturing share through the year 2000, it should be noted that in Japan, the manufacturing share peaked around 1970, when per capita GNP in 1984 US dollars was 7700, and the share of the service sector began to expand. The major policy implication of these projections is the sharp drop in the size of the agricultural sector, which currently accounts for 30% of national employment. A conservative implication of these projections, ignoring technical progress, is that employment in the agricultural sector might fall to half its present share by the year 2000. This raises a clear issue of labor absorption in the nonagricultural sector.

1.19 Table 1.13 shows the projections of the manufacturing structure for years 1990 and 2000, along with the corresponding employment projections. According to KIET estimates, output will continue to shift from light manufacturing to heavy manufacturing between 1983 and the year 2000.[15/] The output composition within heavy manufacturing is expected to change drastically, however. The sectors projected to gain shares are machinery, electronics, automobile and industrial chemicals, while the sectors projected to lose shares are shipbuilding, petrochemicals, petroleum refinery, and iron and steel. Employment shares reflect these anticipated output shifts quite strongly, with the share of employment provided by electronics almost doubling, and by autos more than doubling, while textiles falls by 70% over the period.

13/ This two level approach was necessary because the two sets of projections have been provided by two different institutes. The Korea Development Institute (KDI) published the projections of overall economic structure in Long-term Development Plan for the Year 2000 in September 1985 by incorporating the projections on the manufacturing sector by the Korea Institute for Economics and Technology (KIET). Since then KIET revised the earlier projections and published new projections in October 1985 (Projections of manufacturing structure and development strategy toward 2000). For this reason, the KDI projections are used for the overall economy and the revised KIET projections are used for manufacturing.

14/ Per capita GNP for Japan in 1966 in 1984 US dollars was 5249.5, close to the projected level of 5,016 US dollars in 1984 prices for Korea in 2000.

15/ The KDI projections assume 3% and 6% annual growth rates in the world economy and in world trade (in constant prices), respectively, and 7.4% and 6.1% in annual real GNP and real per capita GNP growth, respectively, between 1985 and 2000. As a result, GNP and per capita GNP in the year 2000 are expected to be $252 billion and $5,103, respectively (in 1984 prices).

Table 1.12: GDP PROJECTIONS
(percentages based on current prices)

	Korea			Japan	
	1984	1990	2000	1966	1982
Shares of GDP					
Agriculture, forestry, fishing	14.0	11.0	7.5	8.8	3.5
Mining and manufacturing	30.4	31.9	33.0	33.5	30.3
Social overhead	18.6	19.2	19.2	16.9	18.5
Services	37.0	37.9	40.4	40.8	47.7
Total GDP	100.0	100.0	100.0	100.0	100.0
Memo Items:					
GDP in 1984 billion US Dollars	81.1	112.0	248.0	523.8	1,395.8
Per capita GDP in 1984 US Dollars	1,999	2,542	5,016	5,249	11,785

Source: Korea: Long-Term Development Plan for the Year 2000 (in Korean), Korea Development Institute, 1985. Japan: Yearbook of National Income Accounts Statistics, UN.

These projections are based on the assumption that the Korean economy will gain a comparative advantage first in capital intensive goods and then in skill intensive goods. This pattern mirrors quite closely the production and export patterns of Japan, which between 1965 and 1981 was transformed from a physical capital-intensive exporter to a human-capital intensive one.[16/] The major differences to be kept in mind, however, are the relative disparities in domestic market size as well as the changing environment for international trade. Neither the domestic nor the external base conditions facing Japan in the 1960s and 1970s is now operative for Korea. Therefore, the implications of the significant industrial transformation being planned merit further examination.

16/ See Noland (1985).

Table 1.13: COMPARISON OF OUTPUT STRUCTURES IN KOREA AND JAPAN
(percent)

	Output			
	Korea /a		Japan /b	
	1983	2000	1965	1983
Machinery	10.4	12.1	12.2	14.3
Electronics	8.2	15.4	9.5	18.0
Automobile	3.6	8.6	2.7	6.3
Shipbuilding	4.3	3.3	3.3	2.6
Petrochemical	3.2	2.6	}	}
Industrial chemicals	3.4	4.6	} 4.6	} 3.1
Petroleum refining	9.9	3.8	}	}
Iron and steel	7.6	6.2	3.3	2.7
Textiles (excl. garments)	13.9	8.3	17.3	10.2
Food	10.9	7.0	6.7	4.3
Other manufacturing	24.6	28.1	40.4	38.5
Total Manufacturing	100.0	100.0	100.0	100.0

Source: KIET: Projections of Manufacturing Structure and Development Strategy Toward 2000, 1985.

1.20 Three basic issues can be identified as being of critical importance as Korea attempts this major industrial transformation. The first relates to employment and concerns not only the overall rate of job creation but also the skill mix requirements. The second issue relates to productivity, particularly total factor productivity, since at some point limitations on further very high rates of increase in the savings rate must be anticipated, and combined with natural limits on labor growth, greater productivity must be squeezed out of Korea's factor inputs. And third, technology will need to be upgraded if Korea is to compete in new industries, and thus a major focus on productive R&D investments is to be anticipated.

Employment Issues

1.21 Employment is already a serious concern to policymakers, as prospects for higher unemployment rates among white-collar workers rise and labor displacement occurs in both structurally declining industries (such as overseas construction) and cyclical depressed industries (such as shipbuilding).[17/] The employment issue is widely expected to become more prominent as: (i) the employment elasticity with respect to output has fallen from an estimated 0.45-0.50 in the 1970s to about 0.25 in the early 1980s, where it is expected to remain; (ii) significant labor displacement in agriculture is expected; and (iii) the skill mix requirements will change substantially as employment opportunities shift from the low-skill to high-skill jobs as a consequence of industrial change.[18/]

1.22 With respect to overall job creation, capital for labor substitution can only proceed so far; however, the increases in employment will occur in skilled areas and in the service sector. Investments in human capital will take on greater importance as the overall technology level of the economy increases. The rural employment problem may prove less tractable, as the rural population may be older, less mobile and less trainable. Government has already given priority to the location of industry, including agro-industry, in rural areas. And with respect to the labor mix, the three industries projected to grow fastest between now and 1996 (viz, electronics, machinery and autos) will increase their share of the current manufacturing work force and will in general be requiring higher skill levels than do the light manufacturing industries.[19/] This brings the area of retraining to light, in which public sector initiatives would be helpful. It also serves to illustrate the potential additional benefits of SMI promotion policies, as smaller firms are more employment-intensive, and the electronics, auto, and machinery components industries are quite suitable for SMIs in terms of subcontracting.

Productivity Issues

1.23 One key for achieving the industrial objectives reflected in the out-year projections is productivity. Total factor productivity (TFP) calculations, as reported in Appendix B of this chapter, show clear drops in TFP

17/ See KDI, "Long-Range Prospects for Manpower Supply and Demand and Policy Tasks," December 1985.

18/ A recent KDI study on manpower policy indicates that projected annual output growth rates for the 1983-96 period will be highest for the electronics industry (12.7%), followed by the machinery industry (10.5%), and transport equipment (9.4%).

19/ KDI and KIET resources.

during the 1970s in Korea.[20] It is the microeconomic level where TFP lapses must be attacked. Government's actions to reduce economic rents and increase domestic competition, through import liberalization for instance, are useful to increase the incentives to innovate and cut costs at the firm level. A second area of importance in reviving TFP is in the area of so-called knowledge factors. Indeed, it is this factor in particular which according to the sources of growth decomposition in Appendix B has performed poorly in recent years in Korea. And this is precisely the factor of production which must be boosted if Korea is to apply scientific technology effectively. Whereas past industrial transformation have been based on superior labor productivity, organizational skills and the like, the current shift being attempted will rely on application of technology and human skills. In that respect R&D and manpower training policies will take on far greater significance than in the past.

R&D Policy

1.24 It can perhaps be said that no area of industrial policy commands greater importance in the eyes of Korean policymakers than R&D policy. R&D investment has increased substantially in recent years, increasing almost fivefold between 1980 and 1984 to W 958 billion in 1984, or 1.46% of GNP. This is still low compared to the OECD countries (see Figure 1.2) but is among the highest for industrializing counties. The mix of funding has shifted from almost entirely public expenditures to rough parity in 1980 and now to a one-quarter/three-quarter sharing with the private sector. This is important because the private sector must ultimately adapt and use technology. While Korea still lags far behind in the basic sciences, and in the production of scientists, it is making great strides in reducing the science and technology gap.[21] Government recognizes the importance of R&D investments, for example, and is aiming to reach 2% of GNP by this year, which would be a doubling since 1983. The challenge, of course, is to successfully apply the new technologies, to adapt foreign technology successfully, and to develop sufficient indigenous capacity to be able to interact successfully with the major new technology producers in Japan and the US.[22] Data from Table 1.14 reveals the extent to which Korea has begun to rival Japan and the US in R&D expenditures.

20/ According to KDI estimates, Korea's annual growth rate of TFP fell to less than 1% during 1970-77 from 3.1% in 1960-69 compared to 4.2% for Japan, 5.7% for Germany, and 2.3% for the U.S. Korea Exchange Bank, 1985.

21/ See Science and Technology Policy in Korea (Ministry of Science and Technology, 1985) and W.Y. Lee (1984) for details.

22/ Is is known, for example, in the semi-conductor field, as reported in the case study on electronics (Volume III of this report), that technology is more likely to be licensed among technological equals.

Shares of R&D Expenditure in GNP (%)

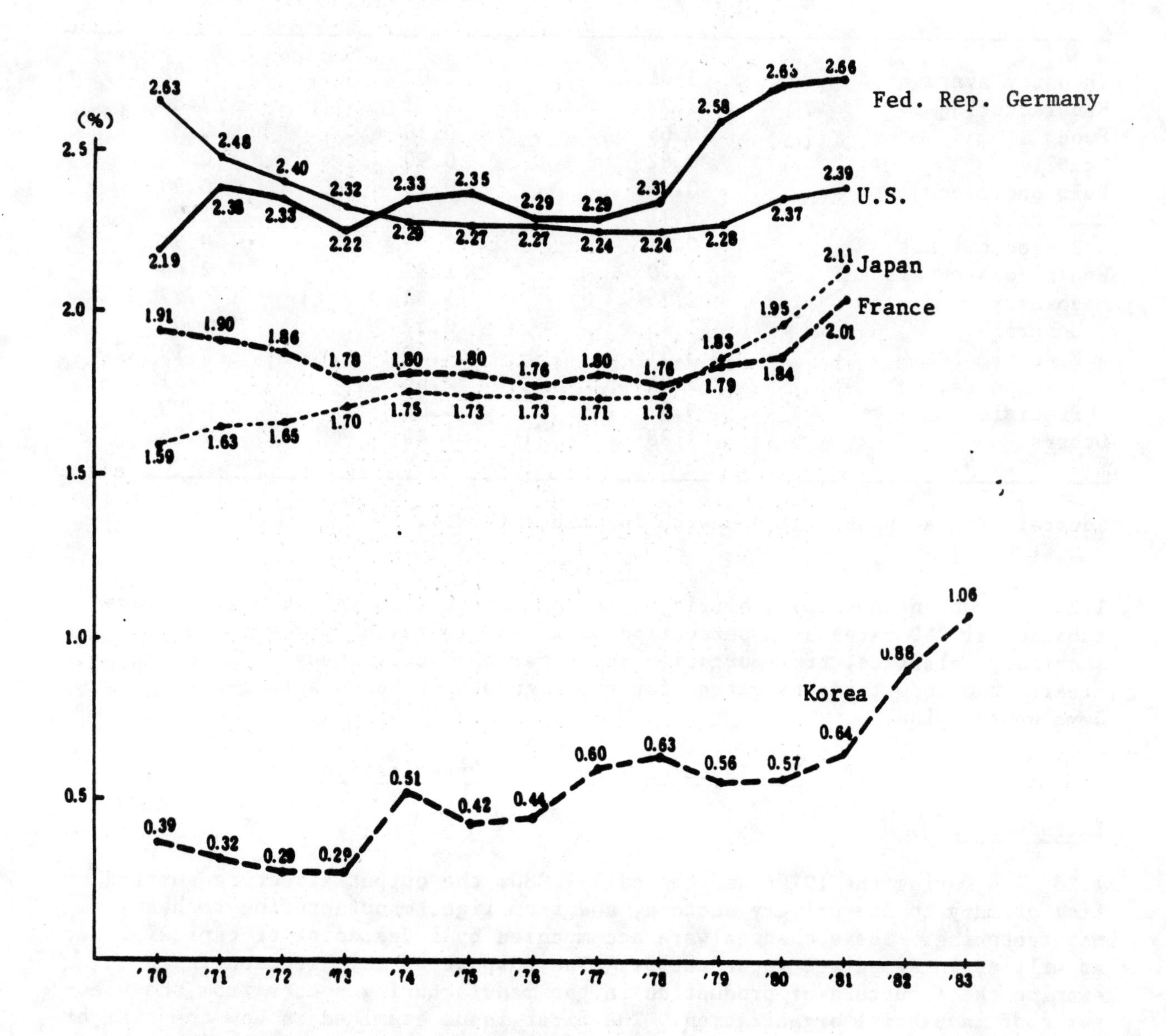

Science and Technology Yearbook, 1984, MOST

Figure 1.2

Table 1.14: COMPARATIVE R&D EXPENDITURE-TO-SALES RATIOS

Industry	Korea (1984)	Japan (1983)	US (1984)
Industry average	1.01	2.03	2.88
Manufacturing	1.27	2.31	2.84
Food	0.65	0.70	0.87
Textiles	0.82	0.90	0.81
Pulp and paper	0.56	0.55	0.98
Chemicals	0.71	1.96	1.94
Petrochemicals	0.31	0.49	0.91
Nonferrous metals	0.99	1.82	2.44
Machinery	2.25	3.38	4.27
General	1.45	2.57	3.53
Electrical	3.51	4.70	5.19
Transportation	1.21	2.66	3.61
Precision	3.07	4.02	5.98
Others	1.38	1.40	2.58

Source: Korea Industrial Research Institute (KIRI).

1.25 On an industry-specific basis, as can be seen in Table 1.15, very substantial R&D rates as a percentage of sales are taking place in the machinery, plastics, transportation and other chemicals areas. These are clearly the industries targetted for rapid growth in Korea's long-term development plan.

E. The Issue of Industrial Organization

Introduction

1.26 During the 1970s and the early 1980s the output structure shifted from primary to non-primary sectors, and from light manufacturing to heavy manufacturing. These changes were accompanied by a deepening of capital-labor as well as intermediate input ratios. The purpose of this section is to examine the structure of production in the manufacturing sector from the viewpoint of industrial organization. The first issue examined is how the size of establishments changed during the 1970s, an aspect of considerable importance in shedding light on the effect of industrialization on the structure of production, and the second relates to how these changes in organizational structure were associated with the changes in the market power of firms and on

Table 1.15: MANUFACTURING FIRMS' R&D

Industries	R&D Expenditure-to-sales ratio 1977	1980		1983	
Food	0.34	0.36		0.70	
Textile and clothing	0.48	0.53		0.73	
Wood products }	0.36	0.34		0.66	
Pulp and paper }				0.65	
Industrial chemicals }				0.58 }	
Other chemicals }				1.62 }	
Petroleum refining }				0.08 }	
Rubber products }	0.78	0.26		0.98 }	(0.56)
Plastic products and miscellaneous }				1.74 }	
Ceramics	1.45	0.52 }		0.77	
Iron and steel }		}		0.28 }	
Nonferrous meatls }	0.52	0.18 }	(0.33)	0.55 }	(0.43)
Fabricated metals }		0.54 }		0.93 }	
Electrical machinery	2.70	1.90		3.01	
Machinery }		1.23 }		2.00 }	
Transport equipment }	0.99	0.62 }	(0.78)	1.48 }	(1.63)
Precision machinery }		1.64 }		1.28 }	
Other manufacturing	2.68	0.16		1.30	
Total	0.95	0.50		0.80	

Source: MOST, Science and Technology Annual.

allocative efficiency in the manufacturing sector.[23/] A preliminary assessment of allocative and technical efficiency is possible by extending the coverage from the firm level to the group or conglomerates level.

Changes in Establishment Size

1.27 Interesting differences are observed in the average size of establishment between the 1973-78 and 1978-83 periods. In terms of employment, for example, the average number of employees per establishment increased from 49.7 in 1973 to 71.8 in 1978, and then declined to 56.4 in 1983. The large increase in the average number of employees per establishment between 1973 and

23/ The distinction between the size measured by establishments and by firms is important, because the former pertains to the production unit while the latter pertains to the marketing unit. Usually there is no one to one correspondence between them because a firm often operates multiple establishments.

1978 had two primary causes. First, the number of establishments employing more than 50 employees increased significantly, as is indicated by an increase in its share from 14.1% to 21.8% (Table 1.16). Second, the average size of large establishments (those with 300 or more employees) increased rapidly from 883.4 employees per establishment in 1973 to 984.2 in 1978. The decline in the average size in 1978-83 was mainly attributable to a rapid increase in the number of small establishments. These patterns mirror official policy towards industry fairly closely insofar as the HCI promotions in the 1970s was accompanied by a shift in the size distribution of establishments to larger firms, and the more recent policy of promoting SMIs in the early 1980s has led to an expansion of the number of small establishments. Using 1983 data, however, two-thirds of output and almost half of manufacturing employment stems from large firms.

1.28 There are clear disparities among industries. Broadly speaking, the nine KSIC 2-digit manufacturing sectors can be classified into three groups according to the importance of large relative to small and medium establishments. The share of large establishments is highest in chemicals, primary metal and metal manufacturing, the HCI sector which is characterized by scale economies, while the share of large establishments is lowest in lumber and wood products, paper and publishing, nonmetallic mineral products, and miscellaneous manufacturing. The remaining two sectors, food, beverages and tobacco, and textiles and leather are intermediate in size. This disparity in the size of establishment is translated into some differences between heavy and light industry with respect to the concentration of output and employment. In HCI, for example, about 2% of the establishments (essentially 400 production units) account for 71% of output and 51% of employment in the sector. Again using 1983 data, comparable figures for light manufacturing show that 2.2% of establishments (about 550 producers) account for 55% of output and 40% of employment. (Figure 1.2 shows the degree of concentration.)

1.29 Compared to the Japanese production structure, the share of large establishments in terms of production activities appears much greater in Korea (see Table 1.17). Although the proportion of large establishments in Korea is less than in Japan, the importance of large establishments in production activities is much greater in Korea. This concentration is important in at least two senses. First it may increase the economy's vulnerability to shocks, insofar as employment in particular is concerned. Second, it demonstrates the extent to which Government may find its hands tied in trying to redirect credit to SMIs while at the same time pusuring ambitious growth objectives, which necessitate strong performances from large production units. The concentration takes on even greater significance in the Korean setting when examined at the firm rather than establishment level because of the high degree of integrated ownership.

Table 1.16: VARIOUS MANUFACTURING PRODUCTION INDICATORS BY FIRM SIZE

Size by (employees)	No. of establishments			No. of employees /a			Gross output /b			Value Added /b		
	1973	1978	1983	1973	1978	1983	1973	1978	1983	1973	1978	1983
5 - 49	20,002 (85.9)	23,328 (78.1)	31,750 (80.9)	236.6 (20.4)	351.0 (16.6)	491.7 (22.2)	363.7 (9.8)	1,907.2 (9.0)	5,793.4 (9.6)	140.3 (10.2)	814.3 (9.9)	2,348.0 (11.2)
50 - 99	1,390 (6.0)	2,818 (9.4)	3,713 (9.5)	97.4 (8.4)	198.0 (9.4)	258.9 (11.7)	247.6 (6.7)	1,441.2 (6.8)	4,524.5 (7.5)	93.2 (6.8)	583.4 (7.1)	1,605.4 (7.7)
100 - 299	1,201 (5.2)	2,581 (8.6)	2,750 (7.0)	205.4 (17.7)	444.0 (21.0)	463.1 (20.9)	642.9 (17.4)	3,511.6 (16.6)	10,713.3 (17.7)	235.8 (17.1)	1,461.3 (17.8)	3,814.4 (18.2)
300 >	700 (3.0)	1,137 (3.8)	1,030 (2.6)	618.4 (53.4)	1,119.0 (53.0)	1,001.5 (45.2)	2,441.0 (66.1)	14,299.5 (67.6)	39,514.5 (65.3)	910.6 (66.0)	5,333.4 (65.1)	13,143.5 (62.9)
Total	23,293	29,864	39,243	1,157.8	2,111.9	2,215.2	3,695.2	21,150.5	60,545.7	1,380.08	8,192.4	20,911.4

/a 1,000 employees.

/b Billion won.

/c Percentage shares are shown in parenthesis.

Source: Major Statistics of Small and Medium Industries 1984, Small and Medium Industry Promotion Corporation. Report on Industrial Census, 1983, EPB.

Table 1.17: ESTABLISHMENTS BY SIZE IN KOREA AND JAPAN: 1982
(%)

	% of establishments		% of employees		% of Value added	
	Korea	Japan	Korea	Japan	Korea	Japan
4 - 99	93.0	96.7	35.6	55.5	19.6	38.6
100 - 299	5.1	2.5	20.4	16.4	18.1	17.3
300	1.9	4.8	44.0	28.1	62.3	44.1

Source: Small and Medium Industry Promotion Corporation, Major Statistics of Small and Medium Industries, 1984.

Economic Planning Board, Report on Industrial Census, 1983, Vols. I and II.

Firm Concentration in Manufacturing

1.30 Firm concentration can be measured as either market or aggregate concentration: market concentration is usually measured as the sales share of largest 3-5 firms in a single market, while aggregate concentration is usually defined as the share of largest 50 or 100 firms in manufacturing sales across markets. The former measure is of interest to Korean policymakers because it measures the degree of market power exercised by a few firms and is thus a public policy concern insofar as protection for consumers is concerned. On the basis of Table 1.18 which compares average "3-firm concentration ratios" for Korea, Japan and a major competitor economy, one can conclude that the market structurc in Korea is the most concentrated. In a different vein, since aggregate concentration reflects the extent of total manufacturing activities controlled by a small number of firms irrespective of the nature of operation, it indicates the extent to which the economy is controlled by the jaebol, Korean conglomerates.

Table 1.18: AVERAGE 3 FIRM CONCENTRATION RATIOS

	All sectors
Korea (1981) /a	62.0
Japan (1980) /b	56.3
Major Competitor Economy (1981) /c	49.2

/a C.H. Yoon and K. Lee (1985).
/b A. Senoo (1983).
/c T.C. Chou (1985).

1.31 The pattern of market structure measured by industrial concentration ratios in Korean manufacturing changed over the course of the 1970-82 period; namely, between 1970 and 1977 in terms of both the number of commodities and the sales value, the share of industries characterized as monopolies increased, while between 1977 and 1982 the importance of oligopoly industries increased markedly, as seen in Table 1.19. The increase in the share of monopoly industries in 1970-77 is probably attributable to the early phase of HCI promotion and a lack of enforcement of any anti-trust actions. Although a decline in the importance of monopolies in 1977-81 is a favorable development, policymakers have to be concerned with a simultaneous decline in the importance of the competitive sales market. The rise of oligopolies in particular points to the prevalence of industrial groups tending to establish dominant positions in a variety of industries, so that conglomerate level concentration is probably even higher than that reflected in firm level data.

1.32 Table 1.20 shows the aggregate concentration for Korea, Japan and a major competitor economy. The shares of sales captured by largest firms in Korea are much higher than in Japan or in the other economy; in 1982, for example, the largest 50 and 100 firms in Korea controlled 37.5 and 46.8% of total manufacturing sales, respectively. In contrast, in a major competitor, corresponding shares were as low as 16.4 and 21.9%, respectively in 1980. Moreover, the aggregate concentration ratio in Korea has been increasing slowly but continously, while the ratios in Japan and the other economy have been essentially declining. The underlying cause may well have been government policy in the area of industrial finance. Historically, public policy has rewarded fast-growing firms with preferential access to credit and industrial survivors have been tapped for expansion. This nurturing of large firms continued during the HCI period and is evident in the bias in industrial finance towards large firms, noted in Chapter 4 of Volume I. The Korean industrial scene is now clearly dominated by very large groups, the jaebol.

Rising Influence of Conglomerates

1.33 Market power was analyzed at the firm level in the previous sections. But this may underestimate the extent of market power exercised in the Korean economy because a substantial number of big firms are controlled by a small number of conglomerates. The importance of conglomerates in the Korean manufacturing sector is seen in Table 1.21. Moreover, conglomerates grew substantially faster than the non-conglomerates as shown by a rapid increase in their sales share, tightening their grip on total manufacturing. The expansion of conglomerates in the 1970s mainly took the form of establishment of new companies rather than acquisition of other companies, while in the 1980s the number of companies belonging to conglomerates declined through mergers and sales.[24/] Thus the size of Korean conglomerates is very large compared to firms in other developing countries is attested to, for example, by the fact that 10 Korean conglomerates are listed among 27 developing

---

24/ The average number of firms in the conglomerates increased rapidly from 4.2 in 1970 to 14.3 in 1979, but then it declined slightly to 13.4 in 1982. See K.U. Lee and J.H. Lee (1985).

Table 1.19: MARKET STRUCTURE OF KOREAN MANUFACTURING
(%)

	1970	1974	1977	1982
Monopoly /a				
Commodities	29.6	30.8	31.6	23.6
Sales	8.7	12.7	16.3	11.4
Duopoly /b				
Commodities	18.7	17.9	20.1	11.1
Sales	16.3	12.6	11.0	6.6
Oligopoly /c				
Commodities	33.2	34.2	32.0	47.4
Sales	35.1	38.6	33.9	50.6
Total Noncompetitive				
Commodities	81.5	82.8	83.7	82.1
Sales	61.1	63.9	61.2	68.6
Competitive /d				
Commodities	18.5	17.2	16.3	17.9
Sales	39.9	36.1	38.8	31.4

/a Firm concentration exceeds 80%.

/b Firms produce more than 80%.

/c Top three firms produce more than 60%.

Source: Yoon and Lee (1985), and Lee (1985).

Table 1.20: SHARES OF LARGEST 50 AND 100 FIRMS IN MANUFACTURING (%)

	Korea		Japan		Major Competitor Economy	
	50	100	50	100	50	100
1970	30.3	40.6				
1972	32.9	43.6				
1973						
1974					16.9 (20.3)	23.4 (28.1)
1975				28.4	15.8 (19.3)	21.7 (26.4)
1977	35.0	44.9			15.2 (18.1)	22.4 (26.8)
1980				27.3	16.4 (20.4)	21.9 (27.4)
1982	37.5	46.8				

Note: For the Competitor Economy, the figures in parentheses indicate the shares of total private manufacturing.

Sources: C. H. Yoon and K. U. Lee, (1985), A. Senoo (1983), T. C. Chou (1985).

country firms in the "Fortune 500" list of the largest non-US industrial companies. However, the size of individual firms in Korean manufacturing is still substantially smaller than those in the developed countries.[25]

1.34 Whereas Government clearly sees advantages in terms of improved international competitiveness in having large, integrated firms, it also is concerned about conglomerate control over financial resources, and has as a result administratively decreed that one-third of banking sector credit be reserved for small and medium firms. This approach may also aim at eventually diluting the power of conglomerates, which, while initially needed to promote Korean exports (i.e., the general trading company concept), have now taken on a life of their own despite the existence of a much enlarged entrepreneurial pool. Contrary to Japan, subcontracting relationships have never really developed in the Korean context, and aggressive forms of vertical integration, as well as horizontal reaching into other product lines has characterized jaebol behavior.

1.35 The jaebol tend to be thinly capitalized, and engage in cut-throat competition, especially when confronted with excess capacity, e.g., the HCI years. More recently, this problem re-emerged in the case of declining industries, such as overseas construction and shipping. Vigorous competition may be seen as adding efficiency and dynamism to the economy, but it also adds additional risk if that competition is conducted under a public umbrella of financial protection. Contrary once again to the case of Japanese business groups (Zaibatsu), which are often reputed to add to the economy's resiliency by internally shifting employment from declining to expanding sectors, Korean firms offer no such implicit employment insurance. An important corollary issue concerns the efficiency of very large economic units.

Table 1.21: CONGLOMERATES SHARE IN MANUFACTURING (%)

	Sales			Employment		
Conglomerates	1978	1980	1982	1978	1980	1982
Top 5	15.7	16.9	22.6	9.5	9.1	8.4
10	21.1	23.8	30.2	13.9	12.8	12.2
20	29.3	31.4	36.6	18.2	17.9	16.0
30	34.1	36.0	40.7	22.2	22.4	18.6

Source: Y.K. Lee (1985).

25/ See Y.K. Lee (1985).

Concentration and Efficiency

1.36 In examining the production and market structure of Korean manufacturing, one may conclude that heavy-chemical industrialization in the 1970s was associated with an increase in the size of production units (establishment size), an increase in market as well as aggregate concentration, and an increase in the size of conglomerates. Unlike the uniform increases in these measures in the 1970s, their development in the 1980s has been more diverse: the size of establishments and market concentration declined, while aggregate and conglomerate concentration continued to increase, reflecting oligopolization at the industry level and a general trend away from competitive structure at the manufacturing-wide level. These findings are of interest because production and market structure have been shown to be related with allocative efficiency.[26/]

1.37 Empirical evidence tends to support the view that the efficiency of small and medium firms caught up with that of large firms by the end of 1970s.[27/] These findings tend to support the government's contention that the SMI sector should be expanded and that it can supply new sources of growth to the economy.[28/] While this is undoubtedly true, and while "affirmative action" policies for SMIs are advantageous in light of the HCI distortions, it is not clear that government should take a position of economic structure on the basis of perceived levels of efficiency. In some industries, in which R&D and large scale technologies dominate, size may be an important determinant of efficiency. In others, minimum critical size may be important to compete internationally. On the other hand, the balance of incentives in the absence of an active SMI promotion policy would clearly favor further agglomeration, which may be undesirable for a number of reasons (see Chapter 5 of Volume I on conglomerates).

26/ Several cross-section studies of the market structure and performance in Korean manufacturing have found that highly concentrated sectors tend to have high profitability. See studies by K.U. Lee (1977), I.B. Choi (1986), and S.S. Lee (1985). There are two radically different interpretations as to the casuality of the positive correlation between concentration and profitability in recent industrial organization literature. The "structuralist view" asserting that the positive correlation is evidence of rent-seeking behavior by firms in oligopolistic industries. It also maintains that larger firms do not have a substantial efficiency advantage over their smaller rivals. On the other hand, the "efficiency-based view" argues that the positive relationship reflects the superior performance of large firms. See, for example, R. Clarke, S. Davies, and Waterson (1984).

27/ KDI estimates show that total factor productivity of SMIs grew considerably faster than that for large firms over the 1970-79 period (4% p.a. compared with 1.4% p.a.) so that they were on par by 1979. See J.W. Kim (1985).

28/ See Appendix 1B on sources of growth.

1.38 In terms of domestic public policy, one clear way of controlling unwanted reductions in competition is through trade liberalizations. While the items subject officially to monopoly oversight will be liberalized at an accelerated pace, there are a wider range of goods produced in oligopolistic markets. Therefore, in the absence of strong antitrust measures,[29] trade liberalization should be aggressively pursued to raise domestic living standards.

29/ There is legislation in the form of Monopoly Regulation and Fair Trade Act (1980) to protect the consumer from unfair practices.

INTERNATIONAL COMPARISONS

A1.01 The similarity of the Korean pattern of change in agricultural and manufacturing GDP shares with international norm calculations,[1/] is instructive, yet there are some striking differences, as seen in Figure A1.1. It should be noted, in particular, that the share of manufacturing increased very sharply in 1972 and it surpassed the international norm in 1973, while the agricultural share remained higher than the international norm throughout the period, perhaps reflective of the high protection accorded to agriculture, as noted in Chapter 5 of Volume I. As a result of these developments, the share of both agriculture and manufacturing was about 2.0 percentage points higher than the corresponding international norms in 1983 (Table A1.1).

A1.02 An important and unique aspect of Korea's structural transformation was the rapid increase in the share of heavy industry during the 1970s, which has been attributed to deliberate government policies. Several reasons are given for the promotion of heavy-chemical industries (HCI) in the 1970s. First, the promotion of HCI was seen to facilitate a shift in Korea's comparative advantage from labor intensive to capital intensive goods in light of various changes taking place at that time, such as rising domestic wage and rising protectionism against unskilled labor intensive goods in developed countries. Second, the threatened withdrawal of American forces from Korea prompted the political leaders to consider rapid establishment of heavy-chemical industries. Finally, the strong aspiration and zeal of the late President Park's regime to achieve the so-called "advanced industrial state" in a relatively short period also played an important role. To promote HCI, various incentive policies were adopted, including special tax treatment, import protection, preferential credit allocation and government investment.[2/]

A1.03 To investigate the effect of HCI push on changes in production structure, Figure A1.2 compares the actual share of heavy-chemical manufacturing in GDP to the corresponding international norm. One interesting finding is the sharp increase in the share of heavy-chemical manufacturing which took place in 1973-1974 compared with the predicted structure. According to the data on credit allocation by sector in Chapter 5 of this volume, the financial preference towards HCI only began to see in 1975, yet the surge in HCI performance compared to international noms had already begun. Moreover, the impetus of the increase in 1973 appears to be different from that in 1974. In 1973, export expansion, due to unprecedented rate of growth in the world trade, contributed a great deal to output expansion of manufacturing while in 1974 it was investment that provided the major impetus to output growth. Between 1972 and 1973 Korean heavy-chemical exports increased by 112.4%, compared to the average annual increase of 63.4% for 1967-1972 while

---

1/ Predicted shares are obtained from regression analysis similar to H.B. Chenery and L. Taylor (1968). For recent extensions of this methodology, see Chenery, Syrquin, and Urata (forthcoming).

2/ For a detailed description, see T. Kwack (1984) and W. Hong (1985).

the corresponding growth rates for light industry were 94.6% and 40.8%, respectively. In 1974 the share of gross capital formation of GNP increased to 31% from 25.7% in 1973. The findings here cast doubt on the conventional wisdom that a shift in the production structure toward heavy-chemical manufacturing was due primarily to the encouragement given by the HCI promotion policies. Instead, the findings suggest that the shift had already started in 1973 and that at least in the early years HCI was export driven as well. This view is further supported by the sources of growth decomposition in Appendix 1B. It should also be noted that the expansion of HCI did not deter the relative international performance of light industry vis-a-vis its norm as shown in Figure A1.3, where the actual share of light industry was significantly higher than the predicted level throughout the period.[3/] Clearly, however, in the counter-factual case, its performance might have been even better, had HCI not received the preferences it did.

A1.04 A comparison of the actual values with the predicted values indicates that the manufacturing share as a whole was substantially higher than the predicted share in 1981 (Table A1.2). It also shows that the actual share is greater than the corresponding predicted share in every sector, except for "other manufacturing" but the structure of production of subsectors within manufacturing is closer to those expected based on level of economic development and size, except for food and beverage sector, where the Korean share is much higher than the predicted share. The high food and beverage share is due to the unusually large agricultural sector, attributable to high domestic agricultural prices. A number of subsectors that exceed expectations are, in fact, in decline (viz., wood and paper) on experiencing excess capacity (viz., chemicals). Comparing the Korean structure with other countries, one finds that the Korean structure is again similar to that of Brazil. Noticable differences are only in textiles and basic metals; Korea has 2.5 percentage point higher share in textiles, while Brazil has 1.9 percentage point higher share in basic metals. Compared to Korea, Hong Kong and Singapore have significantly higher share in textiles and in metal products, respectively. In sum, around 1980 the share of manufacturing as a whole is much higher in Korea than in other countries, but the compositional shares within the manufacturing in Korea are similar to those in other countries.

3/ It should be noted that high protection of primary products is partly responsible for the high share of light manufacturing as the primary products are used as inputs into light manufacturing.

Figure 1.1

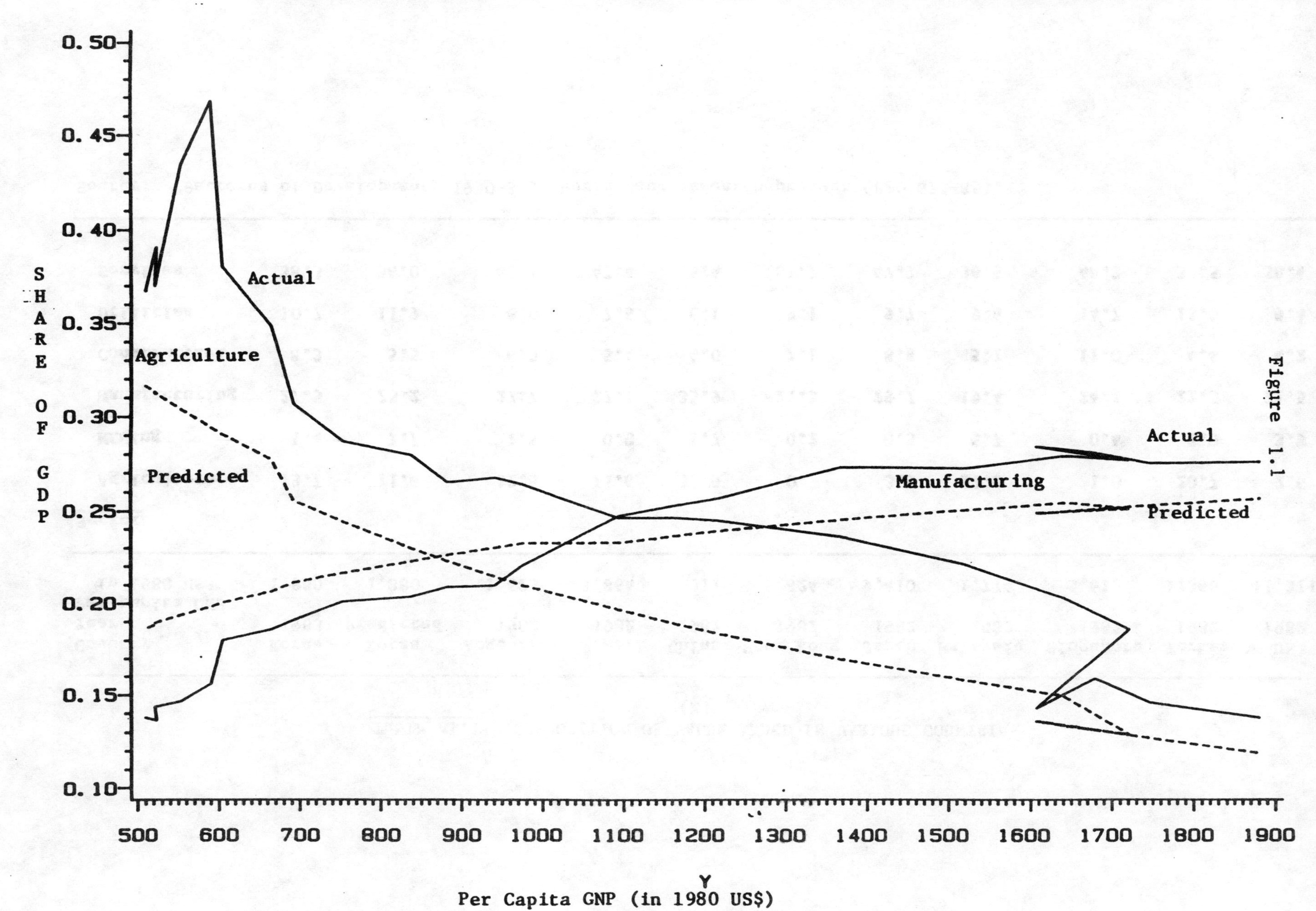
STRUCTURE OF AGRICULTURE AND MANUFACTURING
SHARE OF GDP
0.50
0.45
0.40
0.35
0.30
0.25
0.20
0.15
0.10
Actual
Agriculture
Predicted
Manufacturing
Actual
Predicted
500
600
700
800
900
1000
1100
1200
1300
1400
1500
1600
1700
1800
1900
Y
Per Capita GNP (in 1980 US$)

Table A1.1: COMPOSITION OF VALUE ADDED IN VARIOUS COUNTRIES
(%)

Country	Korea	Korea	Argentina	Brazil	China	Hong Kong	Japan	Malaysia	Singapore	Turkey	US
Year	1983	Predicted	1982	1982	1982	1982	1982	1982	1982	1982	1982
Per capita GNP in 1980 US$	1,880	1,880	1,625	1,864	311	5,824	9,410	1,774	5,613	1,366	11,311
Sector											
Agriculture	13.7	11.6	12.5	11.6	36.9	0.7	3.5	22.6	1.0	20.7	2.6
Mining	1.4	7.7	2.5	0.8	1.7	0.2	0.5	6.7	0.4	2.0	3.9
Manufacturing	27.5	25.2	27.7	27.1	35.9	21.3	29.7	19.4	24.1	22.5	21.5
Construction	8.3	5.5	6.3	5.4	4.0	7.1	8.8	5.1	11.0	4.4	4.2
Utilities	10.7	11.9	8.0	7.6	6.1	9.1	9.7	9.6	14.7	13.0	9.4
Services	38.3	38.0	43.1	47.4	15.4	61.7	47.7	36.6	48.7	37.5	58.4

Source: "Patterns of Development, 1950-83," World Bank research project (RPO 673-85).

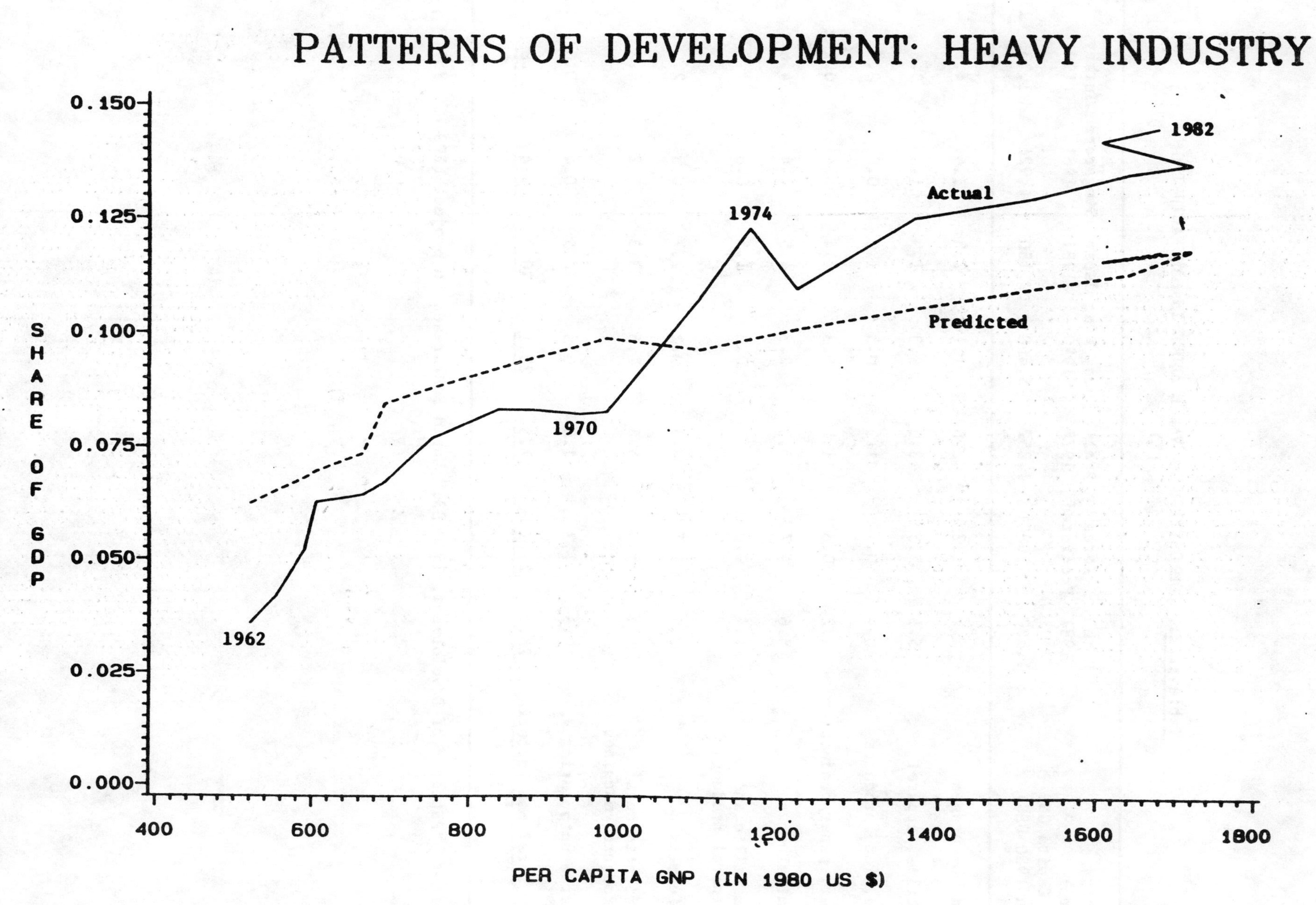
PATTERNS OF DEVELOPMENT: HEAVY INDUSTRY
SHARE OF GDP
0.150
0.125
0.100
0.075
0.050
0.025
0.000
400
600
800
1000
1200
1400
1600
1800
PER CAPITA GNP (IN 1980 US $)
Actual
Predicted
1962
1970
1974
1982

Figure A1.2

Table A1.2: COMPOSITION OF VALUE ADDED IN MANUFACTURING
(%)

Country Year Per Capita GNP in 1980 US$)	Korea 1981 1,680	Korea Predicted 1,680	Brazil 1978 1,852	Hong Kong 1980 5,468	Japan 1981 9,180	Singapore 1981 4,926	Turkey 1981 1,336	US 1980 11,562
Food, beverages	4.4	1.2	3.8	1.1	2.8	1.4	4.6	2.4
Textiles, apparel	5.5	5.3	3.0	9.3	1.9	1.4	3.6	1.4
Wood, wood products	0.4	0.1	1.2	0.3	1.0	0.8	0.3	0.7
Paper, paper products	1.2	0.6	1.5	1.3	2.4	1.4	0.7	2.2
Chemicals	5.6	4.7	5.3	2.3	4.5	7.4	4.6	3.7
Nonmetallic and mineral products	1.4	0.8	1.7	0.2	1.5	0.8	1.9	0.7
Basic metals	2.4	1.7	4.3	0.2	2.7	0.5	2.0	1.4
Fabricated metal products, machines	6.5	5.3	5.8	7.6	13.3	15.2	4.2	10.2
Other manufacturing	0.6	1.3	1.0	0.8	0.5	0.4	0.07	0.4
Total Manufacturing	28.0	21.1	27.5	23.1	30.7	29.4	21.8	23.1

Source: "Patterns of Development, 1950-83," World Bank research project (RPO 673-85).

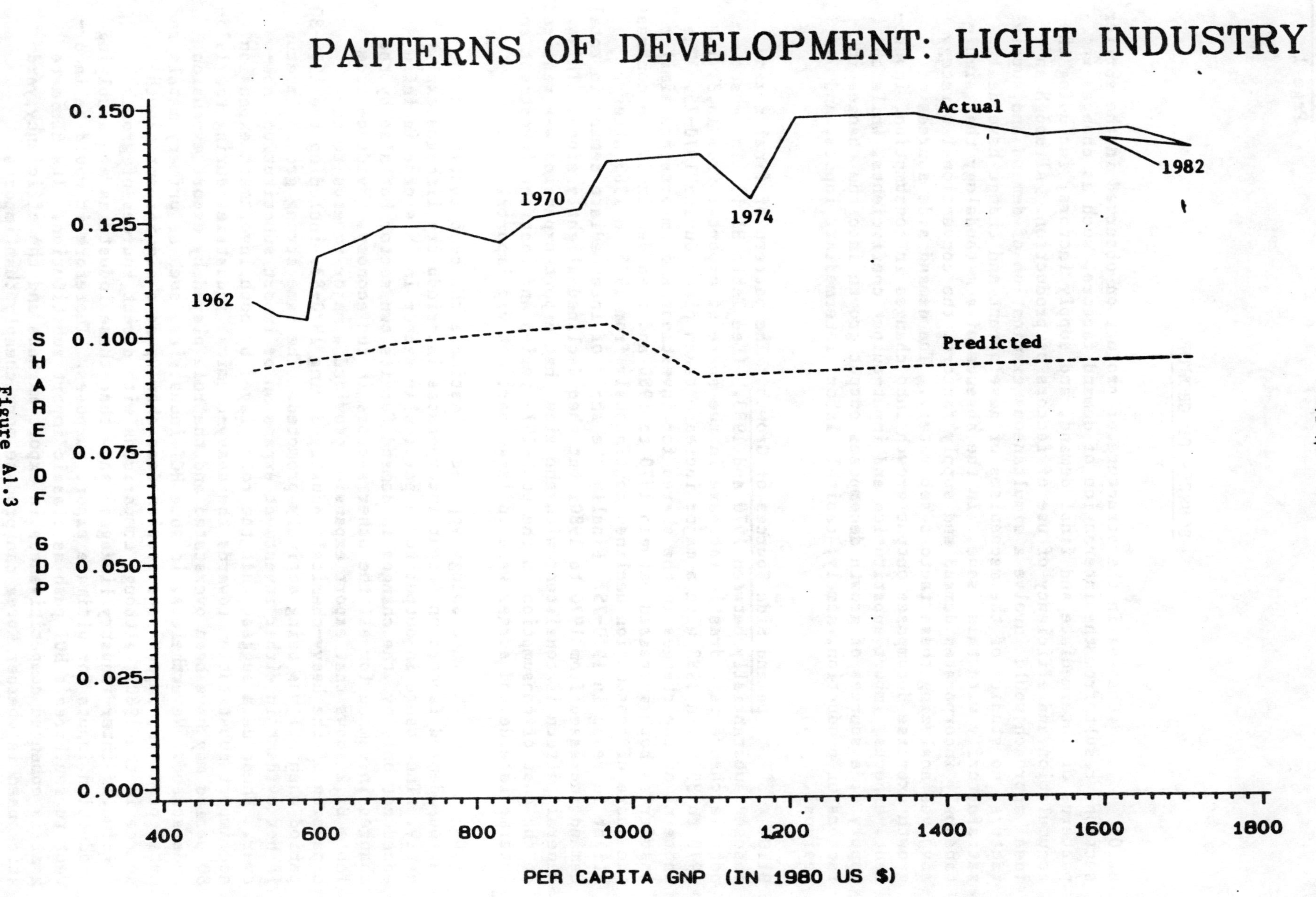
PATTERNS OF DEVELOPMENT: LIGHT INDUSTRY
Actual
Predicted
1962
1970
1974
1982
SHARE OF GDP
0.150
0.125
0.100
0.075
0.050
0.025
0.000
400
600
800
1000
1200
1400
1600
1800
PER CAPITA GNP (IN 1980 US $)

Figure A1.3

SOURCES OF GROWTH

B1.01 Changes in the structure of production observed in the earlier sections result from the interaction of demand factors, such as changes in patterns of intermediate and final demand, and supply factors, including the accumulation and efficiency of use of factors of production. Although an ideal approach would involve a simultaneous examination of demand and supply factors, no studies of the mechanisms of development and change have dealt satisfactorily with this issue. In the absence of a methodology that simultaneously incorporates demand and supply factors, the conventional strategy involves analyzing these factors separately. The demand side sources of growth exercise decomposes output growth into changes in consumption, investment, exports, import substitution and input-output coefficients, while the supply side sources of growth decomposes output growth into the changes in factors of production, namely, capital, labor, intermediate inputs, and residual.

B1.02 Demand Side Sources of Growth. The pattern of final demand changed substantially between 1970 and 1983. (See Table B1.1) The most notable change is a drastic increase in the share of exports: from 13.8% in 1970 to 35.4% in 1983 with a major increase taken place during 1970-75. The patterns of the changes in the shares for investment and imports are similar insofar as both increased between 1970 to 1980 and then declined. In contrast the share of consumption declined continuously from 1975 to 1983 after a slight increase in 1970-75. Finally the share of intermediate demand in total demand increased from 1970 to 1980, but then declined slightly since. This general pattern is consistent with the view that export expansion was realized at the cost of consumption in the post-1975 period and that the investment and intermediate demand surges were in large part met via imports.

B1.03 The methodology applied to estimate the contribution of the various demand factors on output incorporates interindustry linkages explicitly. The change in output for a particular sector in this case is influenced not only by the changes in demand for its own sector but also by the changes in demand for all the other sectors in the economy. Evidence from Table B1.2 shows that export expansion provided a major impetus to the expansion of the heavy-chemical industry during 1970-83, including the 1975-80 period when HCI was being actively promoted. The same strong growth impetus is not evident in either investment expansion or import substitution, however. These data suggest that the role played by both investment expansion and import substitution towards the heavy-chemical industries during the 1975-80 period may have been overstated and the role played by export expansion understated. Nevertheless, in some HCI industries, such as primary metals and metal products, import-substitution does show up as a significant growth source in the 1980s, although comparisons with direct sources of growth (excluding inter-industry linkages) show that these industries are supplying production inputs for ultimate export purposes. Therefore it would be incorrect to see Korea's HCI push as classic import substitution. Its ultimate goal is enhanced competitiveness in export markets and the efficiency yardstick used is whether Korea can produce more cheaply than imports.

Table B1.1: CHANGES IN STRUCTURE OF DEMAND

	Composition of final demand (%)				Growth rate of demand		
	1970	1975	1980	1983	1970-75	1975-80	1980-83
Consumption	84.2	84.7	78.7	74.1	29.1	28.9	15.6
Investment	27.6	26.6	31.5	30.1	28.1	35.2	16.2
Exports	13.8	29.1	33.6	35.4	49.7	34.6	20.1
Imports	-25.6	-40.5	-43.8	-39.6	41.4	32.8	14.2
Total Final Demand	100.0	100.0	100.0	100.0	29.0	30.8	18.0

Source: Bank of Korea, Input-Output Tables.

Table B1.2: SOURCES OF OUTPUT GROWTH IN MANUFACTURING

	Consumption expansion	Investment expansion	Export expansion	Import substitution	Changes in I-O coefficient
1970-75					
Heavy and chemical industry	24.0	23.5	35.9	3.6	13.0
Chemical and chemical products	35.7	9.7	30.6	2.4	21.6
Primary metal manufacturing	8.6	31.1	45.3	8.5	6.6
Metal products and machinery	13.0	41.8	39.8	3.1	2.2
Light industry	57.8	7.9	32.5	-0.8	2.6
Total Industry	54.8	15.5	27.3	-1.3	3.7
1975-80					
Heavy and chemical industry	25.6	22.7	41.3	7.1	3.3
Chemical and chemical products	44.3	7.9	34.9	5.6	7.3
Primary metal manufacturing	7.0	29.8	57.2	12.7	-6.7
Metal products and machinery	9.9	39.9	40.8	5.8	3.6
Light industry	51.7	13.1	31.7	1.4	2.0
Total Industry	48.1	21.2	27.3	1.7	1.8
1980-83					
Heavy and chemical industry	30.5	14.6	54.1	7.9	-7.1
Chemical and chemical products	55.6	15.9	52.8	-3.7	-20.5
Primary metal manufacturing	18.7	-2.7	74.2	19.6	-9.8
Metal products and machinery	17.5	19.2	48.8	12.0	2.6
Light industry	55.6	6.7	37.0	-1.2	2.0
Total Industry	45.8	19.8	32.9	3.7	-2.2

Source: World Bank estimates.

B1.04 Results of demand sources of growth for various periods show that in Korea import substitution was very significant before 1963 (see Table B1.3). In 1963-70 import substitution turned negative and remained negative until 1970-75, reflecting an opening of the economy. From 1975 to 1983, however, the contribution of import substitution reversed sign again, reflecting a positive contribution from import substitution during the period. Export expansion, on the other hand, contributed significantly to output growth throughout 1963-1983. The magnitude of the contribution of export expansion was particularly high during 1970-73, when the Korean exports expanded rapidly thanks to its successful export promotion policies and the expansion in the world trade.

B1.05 An international comparison of the demand sources of growth presents interesting variations in country patterns depending on population size and the trade policies adopted. Large countries (e.g., China, India and Japan) tend to exhibit larger contributions of domestic demand expansion because of their large home market. There are noticeable variations in the importance of foreign trade factors even among the small countries, however. The differences appear to reflect different trade policies pursued in different countries. In Mexico, Turkey and Colombia, all of which more or less adopted import substitution policies during the period, trade factors contributed much less than in other countries of smaller size and endowments. Although export expansion contributed greatly to output increase in Korea, a major competitor economy, Israel, Norway and Yugoslavia, negative import substitution was significant only in Israel, Norway and Yugoslavia, which adopted open economic policies in both export and import markets, and not in Korea or its competitor, where export promotion and import restriction policies were purused simultaneously.

B1.06 <u>Supply Side Sources of Growth</u>. The traditional approach (due to Denison)[1/] to assess the magnitude of various supply factors which contribute to the growth decomposes net output growth of the whole economy to seven supply factors: (i) increase in labor inputs, (ii) increase in capital inputs; and (iii) total factor productivity, which can include, <u>inter alia</u>, economies of scale, advances in knowledge, and miscellaneous determinants. Recent results for the 1963-72 and 1972-82 periods are reproduced in Table B1.4,[2/] showing that labor inputs grew at around 3%, and the increase in labor inputs was the largest single contributor to net output growth in both two periods, accounting for around 35% of the total. The rate of growth of capital increased substantially from 1.14% in 1963-72 to 2.10% in 1972-82, most likely reflecting increased capitalization in HCI and, as a result, its contribution to overall growth rose from 13.9 to 26.2% between the two periods. Total factor productivity (TFP) declined considerably over the period, with the largest decline occurring in the rate of advances in knowledge. It contributed only 3.7% to growth in the later period. Declining TFPs have been

---

1/ See Denison (1974).

2/ See K.S. Kim and K.K. Park (1985).

Table B1.3: SOURCES OF OUTPUT GROWTH FOR VARIOUS ECONOMIES

		Domestic expansion	Export expansion	Import substitution	Changes in input-output coefficient
China	1956-65	107.2	1.9	9.2	-18.3
	1965-75	85.7	5.6	-1.3	10.1
	1975-81	80.3	16.4	-7.9	11.2
India	1959-68	81.0	5.1	7.4	6.6
	1968-73	91.6	6.0	9.3	-7.0
Japan	1955-60	87.9	8.0	-4.1	8.3
	1960-65	90.8	15.1	-2.1	-3.8
	1965-70	82.6	14.9	-3.2	5.7
Korea	1955-63	74.5	10.0	21.4	-6.0
	1963-70	81.8	21.9	-1.8	-1.9
	1970-73	51.9	55.7	-3.2	-4.4
	1970-75*	70.3	27.3	-1.3	3.7
	1975-80*	69.3	27.3	1.7	1.8
	1980-83*	65.7	32.9	3.7	-2.2
Major Competitor Economy	1956-61	54.3	23.9	15.1	6.7
	1961-66	61.3	37.6	-1.1	2.2
	1966-71	52.7	49.5	-0.2	-2.0
Mexico	1950-60	85.5	1.0	5.4	8.1
	1960-70	94.1	3.1	6.0	-3.2
	1970-75	89.8	5.9	-1.3	5.6
Turkey	1953-63	92.3	2.5	1.8	3.3
	1963-68	83.6	4.9	8.3	3.2
	1968-73	86.2	11.8	-1.4	3.5
Colombia	1953-66	76.1	10.0	6.9	7.0
	1966-70	69.5	21.8	6.5	2.1
Israel	1958-65	76.7	25.6	3.1	-5.4
	1965-72	69.1	42.0	-18.9	7.9
Norway	1953-61	60.7	40.4	-10.6	9.5
	1961-69	60.2	49.1	-13.3	4.0
Yugoslavia	1962-66	87.6	23.9	-6.2	-5.3
	1966-72	82.0	30.2	-18.4	6.1

Source: Those without an asterisk are taken from S. Urata, (forthcoming).

Table B1.4: SUPPLY-SIDE SOURCES OF GROWTH
(Annual growth rates)

	Korea			Japan	United States	
	1963-82	1963-72	1972-82	1953-71	1963-73	1973-82
Standardized national income	8.13 (100.0)	8.23 (100.0)	8.02 (100.0)	8.81 (100.0)	3.92/a (100.0)	1.55/a (100.0)
Labor	2.92 (35.9)	2.74 (33.3)	3.04 (37.9)	1.51 (17.1)	1.32 (33.7)	0.66 (42.6)
Capital	1.58 (19.4)	1.14 (13.9)	2.10 (26.2)	2.10 (23.8)	0.85 (21.7)	0.69 (44.5)
Total factor productivity (residual)	3.63 (44.6)	4.35 (52.9)	2.88 (35.9)	5.20 (59.0)	1.75 (44.6)	0.20 (12.9)
Of which: Advances in knowledge	1.09	1.89	0.30			

/a Excludes educational impact on workers, included in TFP for comparability.

Note: Figures in parentheses are precentage shares of total growth in standardized national income.

Sources: K.S. Kim and J.K. Park (1985) and Denison (1985).

noted in the US for recent years, for example, and has prompted a livelier interest in R&D expenditures and other knowledge-augmenting activities.

B1.07 The Korean trend may well be dominated in the 1970s by the HCI experience. Additional KDI studies covering manufacturing industries show sharp TFP drops in the mid-1970s.[3/] Numerous explanations are offered, including overestimation of labor hours worked, lack of competition in domestic product markets, and adjustments to new energy-efficient technologies.[4/] A shift in trade strategy from export promotion in the mid to late 1960s to import substitution in the mid and late 1970s may well have led to a decline in TFP growth. Opening up of the economy in the 1960s due to export promotion policies is believed to have resulted in an improvement in productive efficiency through strong competition from abroad, exploitation of scale economies, and introduction of foreign technology. In contrast import substitution policies, which were adopted to promote heavy-chemical industries in the mid to late 1970s, nurtured inefficiency in production.[5/] Lack of domestic competition might have contributed to slow rate of growth in TFP as well. During the 1970s market concentration increased significantly in the aftermath of HCI episode which encouraged the expansion of firm size. In the absence of significant import liberalization efforts, firms in the concentrated market might have opted not to improve their productive effciency, resulting in low TFP growth. Finally, the oil price increase in 1973-74 forced firms to adopt new energy-saving technology, which is not additionally productive, at least in the short run. This could also produce a decline in TFP growth. Most likely, however, is the overcapacity in manufacturing caused by the ambitious HCI investments of the mid-1970s.[6/] It would be important to know whether the recent liberalization program, and the better HCI performance, have succeeded in turning this around. Clearly, as Korea's productive inputs become increasingly constrained, it will have to turn to more efficiency ways of utilizing those inputs to generate growth.

3/ TFP estimates for 1966-71 versus 1971-75 by Kim and Sohn (1979) show a drop from 3.78% during the former period to 1.07% in the latter. A similar exercise for mining and manufacturing by Rhee (1982) shows a decline in TFP growth from 6.14% in 1967-71 to 2.39% in 1972-76.

4/ See Park and Kim (1985).

5/ Nishimizu and Robinson (1984) support this argument, as it found that TFP growth is positively correlated with export expansion while it is negatively correlated with import substitution by examing their relations in Korea, Japan, Turkey and Yugoslavia, among other countries.

6/ See Westphal (forthcoming).

CHAPTER 2: THE CONDUCT OF MACROECONOMIC POLICY, 1982-85

A. Introduction

2.01 During 1980-82, Korea's principal macroeconomic goal was the achievement of stability and recovery from the difficulties engendered by the overheating of the economy in the late 1970s, the rise in oil prices and interest rates in 1979-80 and the subsequent international recession. In the process, Korea further increased its external debt from about $21 billion in 1979 to over $37 billion by the end of 1982, at an annual average rate of growth of 22%. Believing that this rate of growth of debt exposed the country to unacceptable risks (given the unsettled international financial environment) Korea has sought, since 1982, to achieve an improvement in its external debt position while simultaneously maintaining price stability and a reasonable rate of GNP growth. Specifically, the Fifth Plan (1982-86) aimed at balance of payments equilibrium in the current account together with an inflation rate of under 4% and GNP growth of 7.5%. Macroeconomic policies have had to be chosen carefully to accomplish these goals since policy dilemmas abound when multiple goals are sought.

2.02 Policy Tradeoffs. The notion of a policy tradeoff arises since aggregate demand management policies typically achieve desirable goals (such as the reduction of inflation) at the cost of some undesirable consequences (such as reduced investment and growth and increased unemployment). In Korea the policy environment of the past several years has been characterised essentially by tight money, balanced budgets and exchange rate depreciation (see Table 2.1). Money supply has been kept tight so as not to reignite inflation or put pressure on the balance of payments (BOP). The worry, expressed usually by the business community, is that tight money can have negative effects on investment, exports and employment. Balanced budgets are deemed necessary for price stability and BOP improvement and Korea has pursued tight fiscal policy not simply through nearly balanced budgets but also by keeping the ratio of expenditures to GNP constant. The downside risk to this policy is similar to that for tight monetary policy -- possible adverse consequences for investment, employment and growth. Finally, the won has been steadily depreciated so as to promote the balance of payments objective. The policy dilemma that faced Government was a possible adverse effect on inflation and domestic output (the so-called contractionary devaluation problem), and the possible ineffectiveness of depreciation, in an import-dependent economy, in improving the overall payments position.

Table 2.1: RECENT MACROECONOMIC PERFORMANCE AND POLICY STANCES

	1980	1981	1982	1983	1984	1985
Policy Stance						
Monetary						
NDA growth /a	35	39	34	18	14	24
Interest rate:						
Nominal /b	21	20	12	10	11	12
Real /c	-2.2	4	5	7	7	8
Fiscal						
Deficit/GNP	-3.2	-4.6	-4.4	-1.6	-1.4	1.5
Expenditures/GNP	22.8	24.8	24.2	21.3	21.1	21.4
Exchange rate						
REER change /d	-6.8	2.7	-0.3	-7.3	0.2	-11.0
Macroeconomic Performance /e						
GNP growth	-5.2*	6.6	5.4	11.9	8.4	5.1
GNP deflator	28.6*	15.4	6.6	3.9	3.8	3.6
Current account/GNP	-8.8	-7.0	-3.8	-2.2	-1.7	-1.0
Investment growth	-23.7*	6.3	0.1	17.5	18.6	1.6
Investment/GNP	32.1	30.3	28.6	29.9	31.9	31.2
Savings/GNP	20.8	20.5	20.9	25.3	27.9	28.4
Debt ($ bln)	27.3	32.5	37.1	40.2	42.6	45.6
Debt service (%)	19.7	21.2	22.6	20.9	22.6	23.6
Short-term to total debt (%)	34.3	31.5	33.5	30.1	26.8	25.4
Unemployment rate	5.2	4.5	4.4	4.1	3.8	3.8

/a NDA refers to Net Domestic Assets of banking system.
/b Ceiling lending rate on loans of one-year maturity by commercial banks.
/c Nominal rate adjusted by GNP deflator measure of inflation.
/d Rate of change in the value of the real effective exchange rate (REER).
/e All data unless asterisked is based on the new System of National Accounts and thus is not strictly comparable to pre-1980 data.

Source: Bank of Korea, Economic Statistics Yearbook, and Economic Planning Board data.

2.03 The following sections examine the manner in which these policies have been implemented in recent years, discuss why certain consequences occurred and others did not and indicate how policy effectiveness may be affected by changes taking place in the economy. Four themes emerge. Perhaps the most important theme is that the achievement of macroeconomic objectives in Korea is conditioned strongly by external developments. During 1982-84, favorable external developments such as the decline in the prices of oil and other commodities and the economic recovery in the OECD countries (and especially in the US) enabled Korea to reduce inflation as well as to maintain a high rate of investment and growth despite a conservative domestic macroeconomic policy stance. External developments were also important to Korea's economic performance in 1985. As the US and other OECD economies entered a period of stagnation from mid-1984 on, Korean exports plummetted and both investment and growth faltered.

2.04 The response to this withdrawal of external demand stimulus illustrates a second theme which is that the conduct of macroeconomic policy has been flexible and selective. As the economy faltered in 1985 monetary restraint was moderately relaxed, fiscal expenditures were rearranged without being increased, and depreciation was accelerated. Domestic stimulus was provided selectively for investments in the export sector, a strategy that helped boost aggregate demand while maintaining the impetus towards external balance. Flexibility and selectivity were also present during 1982-84. For example, while the overall macroeconomic policy stance was tight, the private sector was treated more liberally than the public sector.

2.05 A third theme is that the job of demand management was made easier by the simultaneous implementation of supply-side policies. Initiatives taken in the direction of trade and financial liberalization and industrial policy reform began to pay off in the form of higher domestic savings, a pass-through of low international inflation, a reduction of import dependence, and improvements in labor productivity and investment efficiency. While restrained macroeconomic policies have helped curtail inflation, part of the inflation reduction has also come about through such supply side developments. Thus favorable external developments, such as the low rate of inflation in Korea's import supplying countries, combined with lower trade barriers and an increase in the efficiency of import use to keep downward pressure on domestic inflation. Similarly, while depreciation and a tight macroeconomic policy stance have improved the current account and reduced the rate of growth of external debt, part of this outcome is attributable to policies which have raised the domestic savings rate.

2.06 The fourth theme that emerges is that the effectiveness of macroeconomic policy, in particular that of monetary policy, is conditioned by two structural features of the domestic financial system namely, the growth of a non-bank financial institution (NBFI) segment and the high degree of debt-reliance in Korea's corporate sector. The growth of the NBFI segment relative to the primary banking sector has increased the uncertainty regarding the ultimate effect of monetary policy, exercised through the banking system, on investment, output and inflation. This is because monetary developments in the NBFI sector can offset developments in the primary banking sector. During 1983-84, for instance, the authorities sharply reduced the rate of growth of domestic credit through banks and simultaneously raised nominal interest rates. Both of these moves were partially offset by developments in the NBFI sector where the rate of growth of credit did not decline as sharply (and in fact rose during 1984) and real interest rates generally fell. Similarly when a more relaxed stance was adopted in 1985, it was offset to some extent by contraction in the NBFI sector. Moreover, the high degree of debt-reliance in

Korea's corporate sector has constrained the independence of monetary policy as a policy instrument.

B. The Determinants of Investment and Inflation, 1982-85

2.07 The monetary policy that has been followed since 1981 can be generally characterized as tight. The rate of growth of the net domestic assets of the banking system (NDA) was reduced progressively from 34% during 1982 to 18% in 1983 and 14% in 1984. At the same time, nominal bank lending rates were permitted to rise and real lending rates almost doubled between 1981 and the end of 1985 (from 4% to 8%).[1/] During 1985, an emerging recession caused largely by a sharp fall-off in export growth, prompted some monetary relaxation and net domestic assets were permitted to grow at 24%. However, a target of 14% growth in NDA has been set for 1986, which indicates that the 1985 interlude should be considered a temporary deviation from a monetary trajectory that has been, and is intended to remain, essentially restrained.

2.08 The aim of tight monetary policy has been to reduce inflation and the current account deficit. Inflation had been brought down from the unusual heights (for Korea) of 25.6% in 1980 to 15.9% in 1981, while the current account deficit had been reduced from 8.7% to 6.9% of GNP. These levels were still considered unacceptably high and hence monetary policy was set to help achieve further progress on these fronts.[2/]

Monetary Policy and Investment

2.09 An interesting aspect of Korea's recent experience with monetary policy is that while the monetary deceleration (as measured by the rate of growth of real and NDA bank lending rates) has been sharp and while inflation and the current account deficit have been reduced partly by this (and partly

---

1/ It is important to note that the period 1983-85 is being referred to here since nominal interest rate policy changed in this period. Nominal bank lending rate ceilings were reduced sharply during 1982 in the wake of a large drop in inflation because of the concern that very high real interest rates might stall the business investment recovery that seemed to be occurring. By late 1983 inflation appeared to be stabilizing at a low level and Government began a program of financial liberalization under which the nominal lending rate ceiling was adjusted upwards (see Chapter 4 of Volume I for details).

2/ A lower rate of monetary growth and higher real interest rates are thought to reduce the rate of growth of economic activity by reducing investment. As this happens, the rates of growth of incomes and prices decline. In this manner, contractionary monetary policy achieves both lower inflation and a reduced current account deficit since import growth declines as domestic income growth falls and exports rise as domestic prices fall. While the economy is adjusting to the lower rate of expansion of money supply and higher interest rates, investment, employment and growth are typically expected to fall relative to trend.

by exchange rate management as discussed later), this has apparently not come about at the expense of a slowdown in economic activity. In fact, while the monetary deceleration was underway, investment grew steadily, continuing a recovery from a negative growth rate in 1980 (see Table 2.1). As a proportion of GNP, gross domestic investment rose from 28.6% in 1982 to almost 32% in 1984. GNP also grew at an average rate of about 8.5% and unemployment declined from 4.5% in 1981 to 3.8% by the end of 1985. Clearly other factors were at work to offset the potentially adverse consequences of tight monetary policy.

2.10 Three factors can be identified as being important. The first was the pull exerted by the recovery of the OECD economies beginning in late 1982. The second was the fact that the monetary restraint was not as severe in actual experience as appears from movements in the rate of growth of the banking system's NDA and lending rates. Broader measures of liquidity which include the deposit and credit creation activity of nonbank financial intermediaries indicate a deceleration of smaller proportions. Furthermore, real interest rates in the unregulated curb market actually fell somewhat during 1982-84. A third factor was a shift in the distribution of credit from the public to the private sector. These factors are elaborated below.

Investment and Foreign Income Growth

2.11 Beginning in the last quarter of 1982, the OECD countries, and especially the US, began to pull out of the recession in which they had been mired. As a result their demand for imports grew and world trade volume rebounded substantially from the trough of 1981-82. In particular, the sharp growth of import demand in the US proved a boon for Korean exports. The US market normally accounts for over 30% of Korean exports. The elasticity characterizing the response of Korean exports to foreign income is very high (between 3.5 and 4.5) and, as the OECD recovery accelerated (see Table 2.2) the prospects for exports brightened enormously and fixed investment grew rapidly despite the contractionary stance of monetary policy. It is easy to cast this argument in terms of a standard investment function (see Box 2.1). Among the determinants of private fixed investment are not only the interest rate and credit availability, but also current and expected levels of income and profits. Foreign income growth raised current as well as expected income and profits of Korean exporters and thereby led to higher investment.

The Role of NBFIs

2.12 In fact, closer scrutiny of the data indicates that even interest rates and credit availability did not exercise a very strong negative effect on investment. Movements in the banking system's credit supply figures and lending rates do not accurately reflect the overall liquidity situation faced by the private sector to the extent that credit is provided through two other segments, a relatively unregulated segment consisting of NBFIs and a

Table 2.2: GROWTH IN WORLD INCOME AND TRADE

	1981	1982	1983	1984	1985
GDP growth rates (%)					
Industrial countries	1.6	-0.2	2.6	4.9	2.8
USA	2.5	-2.1	3.7	6.8	2.6
Others	1.1	0.9	2.0	3.5	3.0
Japan	4.0	3.3	3.4	5.8	4.4
Korea /a	7.4	5.7	10.9	8.6	5.2
Trade growth rates (%)					
Industrial countries imports (volume)	-2.5	-0.8	4.2	11.7	6.1
Korean exports (volume)	16.4	4.6	15.5	9.8	
Korean exports (value)	20.2	1.0	11.0	13.3	

/a New System of National Accounts methodology.

Source: IMF, World Economic Outlook, October 1985 and Bank of Korea data.

completely unregulated segment, the curb market.[3/] When total credit, including that generated by NBFIs, is examined, it is found to have decelerated much less than the rate of growth of banking system credit (see Table 2.3). In fact, after declining in 1982 and 1983 this measure of liquidity actually rose in 1984. Similarly, when interest rates in the curb market or the yield of commercial paper sold by NBFIs is examined, the trend over 1983-85 is downward (see Table 2.4) in contrast to the trend in bank lending rates. Thus while the government was reducing credit supplied through the banking system, other sources of credit supply did not follow suit. The reason for this is to be found in the particular strategy of financial liberalization followed since 1982.

3/ The primary banking system denotes a set of seven nationwide commercial banks, ten regional or local banks, fifty-two foreign banks and a number of specialized institutions which lend for specific purposes only (e.g., Korea Housing Bank). This is also referred to as the deposit money bank (DMB) system since only these institutions are allowed to generate regular savings deposits. The NBFIs consist essentially of short-term finance and investment companies, and insurance and securities companies. These are permitted to raise cash and lend at rates higher than those permitted to banks. They are also allowed to issue as well as to guarantee bonds. NBFIs have grown significantly in recent years. See Chapter 4 of Volume I for details.

Box 2.1: THE DETERMINANTS OF AGGREGATE PRIVATE INVESTMENT

Private investment is a significant component of aggregate demand in Korea. It is affected not only by domestic macroeconomic policy but also by external developments. Variables representing both of these sets of determinants are included in the investment function estimated below:

$$I = \underset{(2.9)}{22.9} + \underset{(17.5)}{0.75} * I(-1) + \underset{(3.4)}{0.49} * ACC - \underset{(-3.7)}{0.85} * RUM + \underset{(5.3)}{1.91} * DC$$

where the dependent variable, I, represents private real aggregate investment and I(-1) represents investment in the year before. ACC is the accelerator term which is measured as the change in real output during the previous two years; this allows the effects of external developments to be captured indirectly through their effects on domestic incomes and profits. RUM is the real interest rate in the unorganized or curb market and DC is the real flow of domestic credit through the banking system.

The results provide strong statistical support for three claims: (i) increases in real output raise private investment; it is the change and not the level of output that is important; (ii) increases in the cost of capital reduce aggregate investment; and (iii) increases in credit availability increase aggregate investment. These findings have important implications for macroeconomic policy. For example, if financial liberalization increased the cost of capital, there is a risk that investment may suffer. On the other hand, financial liberalization could, by stimulating higher deposits and more efficient intermediation, increase the flow of domestic credit. This would have a positive effect on private investment. Which effect of financial liberalization dominates is ultimately an empirical matter. Experiments with large macroeconomic models tend to indicate that the positive effect will dominate, especially in the long run (see Kwack, 1985).

Estimation details: An instrumental variable technique has been used to predict the curb market interest rate (RUM). Serial correlation in the residuals has been adjusted for (Rho = -0.68). The model is estimated using annual data for the period 1969-83. The numbers noted in parentheses are t-statistics. A t-statistic value greater than 2 indicates that the coefficient is significant at the 95% level. See Edwards (1986).

Table 2.3: MONETARY DEVELOPMENTS
(Annual percentage change)

	1981	1982	1983	1984	1985
Domestic credit A /a	31	25	16	13	18
Domestic credit B	34	30	19	21	23
Domestic credit C	41	39	26	34	29
Domestic credit A					
Private sector credit	26	25	18	14	19
Public sector credit	138	24	-7	-2	2

/a Domestic credit A is domestic credit supplied by monetary institutions while domestic credit B includes credit extended by NBFIs. Domestic credit C measures credit supplied only by NBFIs and development finance institutions such as the Korea Development Bank.

Source: BOK, Economic Statistics Yearbook.

2.13 This strategy had two elements: one was to improve incentives for domestic savings by permitting higher deposit interest rates to be paid and the other was to deregulate the NBFI segment first and permit it to raise deposits and lend at rates higher than those allowed the regular banking segment. As a consequence, total financial deposits have risen since 1982 and NBFIs have captured by far the larger share of the increase. This bias in deposit creation carried through into credit creation (see Table 2.5). Thus the government's financial liberalization strategy has allowed the system to operate with greater liquidity than can be seen by looking at the banking system's liquidity alone.[4]

The Incidence of Credit Reduction

2.14 The manner in which the rate of growth of credit in the banking system was reduced also had implications for investment. The curtailment of credit was borne to a greater extent by the public sector than by the private sector. In fact, in 1983 and 1984 claims on the public sector (government plus public agencies) declined and whatever growth there was in credit supply went to the private sector (see Table 2.3). It can be argued that credit to

---

4/ The same sort of effect may not occur in the future because Government has moved to reduce the bias in favor of NBFIs. Banks have recently been allowed wider spreads and more deposit-attracting instruments such as certificates of deposit. As a consequence their share of financial deposits should stabilize and may even increase.

Table 2.4: INTEREST RATE DEVELOPMENTS, 1982-85

	1982	1983	1984	1985
Selected Lending Rates				
Bank lending rate	12.6	10.0	10.6	11.5
Corporate Bond yield	17.3	14.2	14.1	13.5
Curb market rate	30.6	25.8	24.7	-
Commercial paper yield	19.1	14.4	13.6	13.4
Selected Deposit Rates				
12-Month Deposits				
Banks	8.0	8.0	10.0	10.0
NBFIs	9.0	9.0	10.5	10.5
3-Month Deposits				
Banks	7.6	7.6	6.0	6.0
NBFIs	17.8	13.0	13.0	-
Memo item				
GNP deflator /a	6.7	3.9	3.8	3.6

a/ New SNA methodology.

Source: BOK, Economic Statistics Yearbook.

Table 2.5: STRUCTURAL DEVELOPMENTS IN FINANCIAL SYSTEM

Percent shares in:	1975	1980	1984
Deposit Liability			
Banks	84.1	73.3	57.6
NBFIs	15.9	26.7	42.4
Corporate Finance			
Banks	67.4/a	57.3	35.6
NBFIs	32.6/a	42.2	64.4

/a Refers to 1972-76 average.

Sources: BOK, Economic Statistics Yearbook.

the public sector tends not to have as powerful an effect on investment and growth as credit to the private sector because of differences in investment efficiency. Thus a curtailment of credit which falls mostly on the public sector tends to be less contractionary than the average impact parameter would suggest. The character of investment growth since 1981 reflects this change in the public/private proportions of credit growth. As Table 2.6 shows, while investment has grown as a percentage of GNP since 1982, the share of investment due to government and public enterprises has declined, while that due to private enterprises has increased.

Table 2.6: COMPOSITION OF GROSS DOMESTIC INVESTMENT (percentage of GNP) /a

	1980	1981	1982	1983	1984
Gross domestic investment	31.3	29.2	27.0	27.8	29.8
Private and public enterprises	24.8	22.2	21.1	22.7	24.4
(Public enterprises)			(5.5)	(5.0)	(4.6)
Government	6.5	7.0	5.9	5.1	5.4

/a Based on unrevised SNA methodology.

Source: BOK, Economic Statistics Yearbook, and Economic Planning Board data.

Investment and Interest Rates

2.15 Just as the deceleration in monetary growth in the banking system was partially offset by acceleration in the NBFI segment, so also were higher interest rates in the former partially offset by declining interest rates in the latter. As Table 2.4 shows, bond and commercial paper yields (which reflect the price of credit in the NBFI segment) as well as curb market interest rates declined during 1982-85.[5/] (Interest rate behavior in the non-bank segment is examined in Box 2.2). For present purposes suffice it to note that monetary developments in the primary banking system can be an unreliable guide to predicting real sector consequences.

5/ The yield on commercial paper handled by NBFI's is supposedly determined by the free market and is not regulated by the monetary authorities. The yield on corporate bonds, however, is regulated in the sense that it is not allowed to rise above a ceiling. During 1982-85 these two yields have moved closely together in a downward direction. With the stabilization of inflation in the 3-4% range during 1983-85, these yields have dropped in real terms also.

2.16 The link between private investment and interest rates is important from the point of view of financial liberalization. If interest rate decontrol were to result in a substantial upward movement of the real cost of credit, and if private investment were highly interest elastic, at least in the short-run consequences of financial reform would be strongly contractionary. This has not happened in recent years partly because of offsetting developments in the non-bank (and curb market) segments and partly because of the effects of foreign income growth. It may also have been due to a low elasticity of investment with respect to interest rate changes.

2.17 As far as future policymaking is concerned, however, it should be noted that structural changes appear to be occurring in the business environment which suggest that the interest rate elasticity of aggregate investment may rise in the future. Among these changes is the fact that the proportion of total investment that is undertaken by the private sector, which tends to be more cost conscious, has been rising in recent years (see Table 2.6). Second, the guarantee of adequate credit even when profit margins are low or financial structure shaky, can no longer be taken for granted; in the past Korean business could afford to be relatively complacent about interest rate increases because credit availability was what really mattered. But now that Government has signalled its intention of a following more of a hands-off approach and of allowing credit availability to be determined to a larger extent by individual firm performance and financial structure, interest rates should become a more critical determinant of investment decisions. Third, real curb market rates have fallen in recent years because of the possible unwinding of inflationary expectations; once such expectations are completely unwound the relationship between controlled rates and curb market rates will once again be dominated by domestic liquidity and foreign interest rate (adjusted for devaluation) developments and the conventional transmission mechanism of monetary policy should assert itself in a stronger fashion.

Macro Policy and Inflation

2.18 Inflation reduction has been a prominent aspect of recent macroeconomic performance in Korea. Inflation came down sharply from 15.9% in 1981 to 4% in 1984. To what extent was this due to contractionary macroeconomic policy? It can be argued that the dramatic inflation reduction of recent years was probably due more to favorable developments on the supply side rather than to demand-management policy. Nevertheless, demand management policy may have played an extremely important, but indirect, role by reducing the level of inflationary expectations.

2.19 The major reduction in inflation occurred during 1980-82 (when the inflation rate was brought down from about 29% to about 7%) when monetary and fiscal policy were in fact rather expansionary (see Table 2.1). When macroeconomic policy became relatively tight in 1983 and 1984 inflation was reduced further but by a small amount to an average of about 3.5%. Furthermore, when monetary policy was relaxed in 1985, inflation did not rise. These developments indicate that factors other than domestic macro policy have probably played a more important part in influencing the course of inflation in recent years. Among these are (i) the decline in the rate of growth of wages and improvements in the rate of growth of productivity; (ii) the substantial decline in prices of imports; (iii) adjustments in the use of energy imports and (iv) a reduction in inflationary expectations.

Wages and Productivity

2.20 Nominal wage growth has decelerated sharply in recent years, from a rate of about 15% in 1982 to about 8.5% in 1984-85 (see Table 2.7). One cannot definitively state whether this deceleration was a cause or an effect of declining inflation. However, attempts were made by Government to bring wages down during this period. These attempts took the form of mandated reductions in the rate of growth of public sector wages (including a freeze in 1984) and of moral suasion in the matter of private sector wage settlements. Since 1982 also, productivity growth, as measured by changes in value-added in the manufacturing sector, has been rising. This increase in the rate of productivity growth has probably reduced the pressure on inflation by reducing unit labor costs.

Table 2.7: PRODUCTIVITY AND WAGES IN MANUFACTURING
(Annual percent change)

	1982	1983	1984	1985
Nominal wage growth	14.7	12.2	8.1	9.0
Real wage growth	7.6	9.2	4.1	5.0
Productivity growth/a	-2.0	3.2	12.0	n.a.

/a Refers to index calculated from value-added in manufacturing data (in 1980 prices).

Note: The nominal wage refers to gross monthly earnings. The real wage growth rate is obtained by adjusting the nominal rate using the GNP deflator.

Sources: BOK, Economic Statistics Yearbook.

2.21 Mirroring the increase in productivity is an improvement in investment efficiency as indicated by a decline in the average ICOR characterizing the manufacturing sector. It is clear from Table 2.8 that investment

Box 2.2: THE DETERMINANTS OF CURB MARKET INTEREST RATES

In a semi-open financial system and economy such as Korea's, both external and internal policies and events can be expected to affect domestic financial markets. Among the factors expected to determine the curb market interest rate are foreign interest rates and expected depreciation (which determine the effective cost of foreign credit), the degree of excess liquidity in the domestic economy, and inflationary expectations. These determinants are included in the following regression model:

$$i = \underset{(1.8)}{0.47} + \underset{(2.0)}{0.08}*(iw+d) + \underset{(3.0)}{0.47}*i(-1) - \underset{(2.6)}{0.08}*\log DC(-1)$$

$$+ \underset{(0.3)}{0.08}*\log Y + \underset{(1.8)}{0.38}*TDR$$

The dependent variable i is the nominal curb market interest rate. The variable iw is the US Treasury Bill rate and the variable d measures the actual rate (a proxy for the expected rate) of devaluation of the Won with respect to the US dollar. DC(-1) is the amount of domestic credit supplied by the banking system in the previous period. Y is the level of real GDP and is a proxy for the demand for money. Together, DC and Y provide an indication of the degree of excess liquidity in the economy. TDR is the term deposit rate for one-year bank deposits; this variable adds further information about the state of domestic liquidity and about the opportunity cost of borrowing from the curb market. As far as inflationary expectations are concerned they should presumably be reflected in the expected rate of devaluation.

The statistical results shown above indicate that the curb market interest rate is indeed influenced by both external and domestic liquidity factors. From a policy point of view, the results suggest that nominal devaluation can raise the domestic cost of credit. If the latter is an important component of unit costs, inflation could be increased and/or output decreased. However, the effect detected here is quantitatively rather weak. The results also suggest that curb market interest rates tend to rise when controlled interest rates in the primary banking system are permitted to rise and to fall when credit supply in the primary banking system is increased. Thus, government monetary policy can affect developments in the unregulated segment of the financial sector.

Estimation details: The model was estimated by the method of ordinary least squares using quarterly data for the period 1977-85. The R-squared statistic was 0.88 and the Durbin-Watson statistic was 1.38. The numbers in parentheses are t-statistics. A t-statistic value greater than 2 indicates that the coefficient is significant at the 95% whereas one greater than 1.6 indicates significance at the 90% level. See Edwards (1986).

efficiency has begun to increase after deteriorating sharply in the late 1970s.[6/] The improvement can be attributed to the decline in the proportion of investment originating in the public sector or at the behest of Government through directed credit arrangements. In particular, the improvement in efficiency can be attributed to the reduction of the rate of growth of investment in the HCI sector and an increase in capacity utilization in this sector. It is also possible that some of the improvements in efficiency and productivity have come about because of intersectoral shifts of resources from low productivity to high productivity uses, from the growth of skills in the labor force and from technological improvements arising from indigenous R&D or imports. While it is difficult to quantify the contributions of each of these factors, there is evidence which suggests that they have indeed been important.[7/]

Table 2.8: AVERAGE INCREMENTAL CAPITAL OUTPUT RATIOS, 1978-84 /a

	1978-81	1982-84
Agriculture, forestry & fishing	6.13	3.6
Manufacturing	5.33	1.6
Construction	2.69	1.0
Wholesale & retail trade, restaurants & hotels	7.96	2.6
Transport, storage & communications	13.25	1.6
Financing, insurance, real estate & business services	1.01	0.7

/a Changes in output are lagged by one year. Both output and fixed investment are in 1980 constant prices.

Source: Economic Planning Board and Bank of Korea.

6/ It should be noted that ICOR's are a crude measure of investment efficiency. Nevertheless, in the absence of more refined measures, ICOR's can convey useful information. In particular, trends in ICOR values over time are typically consistent with those of more elaborate measures of efficiency.

7/ For example, the supply of skilled manpower is steadily increasing partly because of increases in university enrolments and partly because of on-the-job training imparted in the new industries that Korea moved into in the late 1970s. The proportion of government and private resources devoted to R&D activities has risen since 1980 and this may have provided some technological improvements, but a more important source of such improvements has probably been imports. Intra-industry trade is becoming important in Korea's trade structure and trade liberalization promises larger technology improvement bonuses over time.

Import Price Developments

2.22 The softness that has characterized the prices of Korea's main imports (such as oil, intermediate goods) has probably helped significantly in the control of inflation since imports are a large fraction of Korea's GNP and import prices typically account for 40% of the wholesale price index (WPI). The import price index has declined from a level of 104 in 1981 to 94.4 in 1984 as oil and commodity prices have slackened. This constitutes a rather significant improvement and given the weight of imports in the WPI it is not surprising that domestic inflation has eased considerably during 1982-84. (See Table 2.9.) These developments continued into 1985 and show every sign of continuing to be important in determining domestic inflation in 1986.

2.23 The fact that such favorable external price developments have been allowed to pass-through into the domestic economy by way of a general lowering of trade barriers illustrates a benefit of the trade liberalization process that Korea has embarked upon. This pass-through would have had even more of a restraining effect on domestic inflation had it not been for the fact that considerations of export competitiveness prompted exchange rate depreciation during 1983-85, and especially in 1985.

Energy Use Adjustments and Inflation

2.24 The control of inflation since 1980 has also been aided by the substantial adjustments achieved by Korea in the level and pattern of energy use. If investment in the 1980s had been based on the pattern of energy use established in the 1970s, a high rate of inflation would undoubtedly have occurred as a rising energy coefficient (ratio of energy use to GDP) would have exaggerated the effects of the sharp rise in the price of oil, the economy's main energy input, in 1979-80. Even when the relative price of oil began to decline after 1982, rising energy use (in volume terms) could have offset this effect.

Table 2.9: IMPORT PRICE INDEX DEVELOPMENTS
(Weights in parentheses)

	1980	1981	1982	1983	1984	1985
All commodities	100	104.0	98.7	94.4	94.4	90.7
Mineral products (33.1)	100	112.0	108.6	96.4	92.3	91.7
Agriculture and related (20.6)	100	97.8	81.9	82.2	86.1	75.0
Industrial products (46.3)	100	101.0	99.1	98.4	99.6	96.9

Source: BOK, Economic Statistics Yearbook.

2.25 There are several reasons why the energy coefficient could have been expected to rise in the 1980s. Korea's industrial structure changed during the 1970s in the direction of greater energy intensity as such high-energy-consuming sectors as heavy and chemical industries were fostered. Also, the mechanization of agriculture was more or less completed in the 1970s and further investment in agriculture would have had to contend with higher levels of energy use. Finally, real incomes had risen substantially in the 1970s and the consumption of energy intensive durables such as household appliances and cars had begun to rise. Thus, going into the 1980s, Korea could have expected to see a rise in the ratio of its energy consumption to GDP.

2.26 As a result of strong conservation and substitution measures, however, this did not happen. As Table 2.10 shows the ratio actually declined from 1980 to 1983. The principal measures employed were the continuation of a fuel substitution program (begun in the 1970s), which featured the building of nuclear power plants and the passing along of high oil prices to final consumers to reduce oil consumption. Devaluation and surcharges made the retail price of oil even higher than that imposed by OPEC. The elasticity of oil consumption with respect to the GDP fell from 1.2 during 1973-78 to 0.45 during 1978-84 and the elasticity of total energy consumption with respect to manufacturing output (in real value added terms) declined from 0.85 in 1973-78 to 0.37 in 1978-84.

Table 2.10: INDICATORS OF ENERGY USE

	1980	1981	1982	1983
Energy/GDP ratio				
TOE/$'000 /a	0.77	0.75	0.71	0.70
Index (1980 = 100)	100	97.4	92.3	91.5

/a TOE = ton of oil equivalent.

Inflationary Expectations

2.27 Finally, the role of inflationary expectations should also be mentioned. The 1970s were a period of relatively high inflation and government attitude towards investment and growth fed inflationary expections. The process came to a head in 1980 when inflation soared to 25.6%, while real GNP actually declined (by 5.2%) for the first time in Korea's modern economic history. The reorientation of government attitude since then and the steadiness with which the government has pursued a price stability objective has reduced inflation considerably and must have reduced inflationary expectations. Korea has never enjoyed inflation rates so low and for so long as during the past four years. The ensuing decline in inflationary expectations has been of great importance in reducing inflation and has probably itself been affected strongly by the governments aggregate demand policy stance.

While the actual reduction of demand stimulus as a consequence of conservative policy may not have been large, the psychological effect on inflationary expectations has probably been quite strong.

C. Monetary Policy in the Medium Run

2.28 The conduct and effectiveness of monetary policy will be influenced over the medium run by some structural aspects of the Korean financial system. Among these are recent shifts in the structure of the financial system resulting in an enhanced role for NBFI's, and a pattern of corporate finance featuring high debt-equity ratios. In addition, the conduct of monetary policy will be influenced by the goal, emphasized in the Sixth Plan, of maintaining systematic surpluses on the current account.

Monetary Policy and NBFIs

2.29 The structure of the Korean financial system has changed in several respects over the last ten years or so. NBFI's now attract a substantial proportion of household savings deposits and generate a substantial proportion of domestic credit to the corporate sector. They have garnered by far the largest chunk of incremental asset growth in the financial system since 1973. If their role in the issuance and guarantee of commercial bills and corporate bonds is also taken into account, it would be fair to say that they have become as important as the established deposit money banks (DMBs) in the overall financial intermediation process in Korea. This increase in their relative importance has come about at the expense of both DMB's and the curb market. The shifts in asset shares between DMBs and NBFIs has already been documented (see Table 2.5).

2.30 A major implication for monetary policy of the growth of NBFIs is that this complicates the business of setting monetary targets and achieving desired real sector effects. The monetary authorities have more control over the credit creation and credit pricing behavior of DMBs than they do over NBFIs because the latter are subject to less regulation and do not rely on the Bank of Korea for rediscounting or deposit enhancement. As a consequence, when the authorities fix a domestic credit supply target through the DMB system, they cannot be sure what effect this will have on overall liquidity in the economy. As shown earlier (see Table 2.3), overall liquidity can differ both in magnitude and direction of change from DMB liquidity. For example, while the rate of growth of dometic credit supplied through DMBs was reduced from 16% in 1983 to 13% in 1984, that of credit supplied through NBFI's accelerated from 26% to 34%. As a result, overall liquidity growth accelerated from 19% in 1983 to 21% in 1984. Furthermore, when the rate of growth of domestic credit supplied through DMB's was accelerated to 18% in 1985, that supplied through NBFI's decelerated to 29%. The net result was a modest acceleration in overall liquidity to 22.5%.

2.31 These variations in the rates of growth of alternative monetary aggregates indicate the existence of potentially powerful offsets between credit supply behavior in these two market segments. Sufficient experience has not yet been gained through which to specify a rule of thumb that would allow the authorities to predict the overall liquidity consequences of credit supply policies framed with respect to the primary banking sector alone. This adds further uncertainty to the matter of predicting real sector consequences. Korea's experience is not unique in this respect. As new financial instruments have proliferated in the OECD economies and as financial deregulation has spread, overall liquidity positions have become less subject to government control. To the extent that this is so, uncertainty regarding the ultimate effect of government policy on investment, output and inflation is increased.[8/]

2.32 It is also possible that the relative growth of NBFIs will alter the resource allocation effects of monetary policy. The firms, sectors and activities that draw largely upon the DMBs for credit will be more strongly affected by monetary policy changes/initiatives than those who draw credit largely from NBFIs. This effect, of course, will be smaller the greater the extent to which these segments have overlapping clients. While large firms can switch more easily between these two segments (indeed, many NBFIs are owned by large firms), small firms cannot. Since the sectoral pattern of access to the different segments of the financial system is not known, Government may wish to investigate the potential bias in monetary policy on small and medium industries versus large firms.

Monetary Policy and the Corporate Finance Structure

2.33 The structure of corporate finance is perhaps the Achilles heel of the Korean economy. Debt-equity ratios for Korean manufacturing firms (see Table 2.11) tend to be high in comparison with those for firms in most developed countries as well as in the NICs.[9/] As a consequence, the financial viability of Korean firms is threatened whenever interest rates rise or credit supply is curtailed by amounts that would be considered quite manageable in other countries. This constrains monetary policy and the process of financial liberalization. Government has had to proceed gingerly with respect to financial liberalization in recent years because of the implicit threat of financial distress among Korea's highly leveraged firms, and especially some large conglomerates which are disproportionately important to output, employment and export performance. The structure of corporate finance will

---

8/ This need not imply that the broadest measure of liquidity is the most appropriate indicator for macroeconomic performance prediction. Narrower measures may still remain appropriate for specific purposes.

9/ Debt-equity ratios have historically been high in Korea essentially because of government-mandated cheap debt and expensive equity policies. Real interest rates have typically been very low or negative while equity has been discouraged through a variety of disincentives such as the prohibition of new issues at prices above par. Government has recently taken steps to improve corporate debt-equity ratios by raising the cost of debt and reducing disincentives to the issue of equities. While some success has been achieved, debt equity ratios remain high and Korean corporations remain hostage to their financial structures.

continue to affect monetary policy in the future until such time as when the excessively heavy reliance on debt is reduced to international norms. Paradoxically, tight monetary policy featuring high interest rates and lower levels of credit supply is one of the ways in which to discourage debt reliance. This is best implemented, however, when business conditions are good and firms have higher levels of profits with which to finance credit needs and/or can issue equity at attractive prices.

Table 2.11: COMPARATIVE DEBT-EQUITY RATIOS (MANUFACTURING SECTOR)

	1980	1981	1982	1983	1984
Korea	488	452	386	360	343
Japan	385	378	392	324	na
US	101	105	106	104	na
Germany	214	222	216	218	na

Source: BOK, Financial Statements Analysis.

Sterilization and Current Account Surpluses

2.34 As mentioned earlier, the most prominent macroeconomic goal of the Sixth Plan is the achievement of systematic surpluses on the current account. Normally this would have implied a continuation of the aggressive, competitiveness-oriented nominal exchange rate policy that Korea has successfully followed in recent years. The sudden and sharp decline in the price of oil and the sharp appreciation of the yen relative to the won in early 1986 reduce the necessity for such a policy. These favorable external developments are likely to produce current account surpluses directly. The implications for monetary policy are not as clear-cut. On the one hand, the downside inflation and external deficit risk to looser monetary policy is now considerably lower. On the other hand, such surpluses will tend to result in exchange rate appreciation and eventually propel the exchange rate to a level consistent with external balance. If, however, Korea seeks to generate systematic surpluses in order to reduce its net debt it will have to be ready to sterilize the foreign exchange liquidity inflows associated with such surpluses.

2.35 Even if Government possesses the will to sterilize it may not have the ability to do so. Successful sterilization requires monetary instruments that are widely acceptable (such as Treasury securities in the US) and formal financial markets that are the predominant channel of the flow of funds in the economy. Korea is at present lacking on both counts. There does not exist a government security that the general public is willing to hold in meaningful quantities. Stabilization and sterilization are presently conducted through the forced sale of government securities to commercial banks. To the extent also that financial markets are segmented and commercial banks do not control the bulk of the flow of funds in the economy, sterilization operations may be neutralized by offsetting movements in the nonbank segment. If the capital account is open there is an additional source of offsetting liquidity in the form of international capital flows.[10/]

2.36 The development of a strong sterilization instrument is linked to the process of financial liberalization. The main reason why government securities are not widely held is that Goverment has found it more convenient to "borrow" from pliant commercial banks at essentially zero interest rates. Government could spread the distribution of its securities among the public by offering interest rates competitive with those available to the public in the nonbank segment of the financial system. Once all interest rates in the financial system are allowed to be market-determined, the system will become unified and once government control over the asset management of commercial banks is reduced Government too will have to borrow at market rates of interest. Both of these outcomes will strengthen the sterilization ability of the government.

D. Exchange Rate Management

2.37 In recent years the exchange rate has featured prominently among the macroeconomic instruments employed to stabilize the economy and to adjust to the economic change wrought by the events, already alluded to, of 1978-81. The stabilization goal has essentially been to avoid large current account deficits. The adjustment goal has had two aspects: one to adapt to the changing structure of international prices (especially for oil) and interest rates and the other to facilitate the liberalization of imports (in order to benefit from gains from trade and transfer of technology) and to open up Korea's domestic market without serious dislocation. Towards these ends, the nominal exchange rate has been depreciated in small steps (after a devaluation of 17% in January 1980). The nominal depreciation has been accompanied, for the most part, by a real depreciation. According to one index, Korea's real

---

10/ The empirical evidence on the Bank of Korea's ability to sterilize is mixed. The results tend to range from suggesting complete inability to control monetary aggregates to a rather strong ability to do so in the short run. The difficulty of interpreting the existing empirical studies is compounded by the fact that the policy environment, which reflects the will to sterilize, has not remained constant and hence the selection of different periods for study can yield different results.

effective exchange rate has depreciated by almost 20% since 1982.[11/] The results of this exchange rate management have been successful so far and Government is expected to continue to use the exchange rate to attain both short-run macroeconomic objectives and long-run structural ones.

Impact on the Current Account

2.38 One measure of this policy's success has been the taming of the current account deficit; this has been brought down from 6.9% of GNP in 1982 to about 1% in 1985. The success of depreciation in boosting export and restraining imports has been due in considerable measure to the price elasticities that characterize Korean tradeables: a 10% decrease in export prices raises exports by almost 17%, while a similar increase in import prices causes imports to fall by about 5% (see Table 2.12).[12/] One should avoid the

Table 2.12: RANGE OF ESTIMATES OF KEY TRADE ELASTICITIES

	Earlier period (1962-78)	Recent period (1974-84)
Income elasticity of imports	1.4 to 2.3	1.0 to 1.3
Import price elasticity of imports	1.4 to 1.7	0.4 to 0.6
Foreign income elasticity of exports	5.5 to 6.4	3.5 to 4.3
Export price elasticity of exports	1.4 to 3.2	1.4 to 1.7

Sources: Kwack (1985); Kwack-Mered (1980); Y.B. Kim (1984).

11/ While the nominal exchange rate is the policy instrument available to the authorities it is the real effective exchange rate which actually determines ultimate effects on the balance of payments and resource allocation between tradeables and non-tradeables. This rate is affected partly by nominal devaluation, partly by domestic macroeconomic policy, and partly by external developments (see Box 2.3). In this sense, exchange rate management is not entirely independent of macroeconomic policy.

12/ Korea's success with exchange rate depreciation in recent years should not come as a surprise to students of Korea's macroeconomic history. Two episodes are worth recalling. A devaluation of 18% (against the dollar) in late 1974 produced a spectacular recovery of the export growth rate in 1975 even though the volume of world trade (in manufactures) fell that year by about 5%. A similar development occurred in 1980 when a devaluation of 17% (against the dollar) helped produce an export growth surge in 1980-81. The general policy of depreciation to enhance external competitiveness and control the current account deficit has been more or less consistently followed since the devaluation of 1980. The overall effects of a nominal depreciation are estimated in a number of large macroeconomic models for Korea (see Kwack, 1985, and BOK, 1985). It is generally confirmed that such a policy can improve the current account.

temptation of extrapolating into the future from these elasticities, however, because a larger proportion of Korean exports are covered by increasingly tight protectionist arrangements now (and will be through the foreseeable future) than has been the case in the past. In addition to textiles and clothing, the traditional Korean exports, newer Korean exports such as steel products are now subject to quotas in OECD markets. The greater the extent to which such important exports are controlled by quota arrangements, the less the advantage to be gained by price reductions. In fact a different incentive may apply under quantitative restrictions to the extent that higher value-added, higher price items should be exported so as to maximize the value of the quota to the exporter. It should also be emphasized that Korea faces intense competition from other NIC's and aspiring NIC's in its major export markets. Depreciation of the won is unlikely to be successful in expanding market shares if competitors follow suit and depreciate their currencies proportionately. This aspect of increasing competition should be kept in mind by Korean policymakers when assessing the future role of exchange rate management.

2.39 On the import side, however, there may be increasing room for maneuver. This is largely a consequence of changes in investment mix and of adjustment to higher oil prices in recent years. As Government has gradually abandoned its promotion of heavy industries, which were highly import-intensive, the import-dependence of Korea's exports, investment and consumption goods has begun to decline (see Table 2.13). Second, the development of the capital goods industry in the 1970s is now paying off. Finally, conservation and other responses to high oil prices have made Korea less dependent on oil imports (on a per unit output basis) than before (see Table 2.10).[13/] Nevertheless, at least in the short-run, sharp movements of the won vis-a-vis the yen, for example, will not provide an immediate and sizeable payments improvement to the extent that Korean exports are dependent on intermediate inputs which cannot easily be sourced.

2.40 Exchange rate management to promote improvement in the current account has been helped by conservative macroeconomic policy in recent years. The literature on the effects of devaluation is fairly unanimous in indicating that expansionary macroeconomic policy tends to cause an appreciation of the real effective exchange rate (REER). High ratios of public expenditure to GNP and high rates of growth of domestic credit tend to be associated with higher rates of inflation and, if they occur after a nominal devaluation, they often offset the relative price effects of the devaluation. Empirical analysis for the Korean case (see Box 2.3) confirms this. The implication is that the conservative monetary and fiscal policy that has been implemented in recent years

---

13/ The share of petroleum products has declined from 27% of all imports in 1980 to 21% in 1984. This has been partly due to declining oil prices and partly to strong conservation measures.

has probably been very important in maintaining export competitiveness through downward pressure on the real effective exchange rate.[14]

Table 2.13: IMPORT GENERATION COEFFICIENTS

	1970	1975	1980	1983
Consumption	0.13	0.19	0.23	0.22
Investment	0.39	0.48	0.42	0.35
Exports	0.26	0.36	0.38	0.36
Final Demand	0.20	0.29	0.30	0.28

Notes: The import generation coefficient is the value of imports generated to satisfy one unit of final demand. It captures both the direct and indirect requirements for imported intermediate goods.

Source: World Bank estimates.

Impact on Inflation

2.41 Another measure of the success of depreciation in recent years is the concurrent achievement of price stability and reasonably high growth. No serious stagflationary effects have followed the fairly steady depreciation undertaken since 1980. The lack of any visible effect of recent depreciation on inflation is striking because most empirical studies confirm the existence of a positive relationship between domestic inflation and the rate of depreciation.[15] The principal offsetting factor must have been the decline in the import price index (measured in dollars). The moderation of the devaluation--inflation connection may also be due to two developments mentioned earlier, viz., the decline in import dependence in general and the

---

14/ It is possible also that tight fiscal policy reduces import demand in addition to reducing domestic inflation. If government expenditures have a high import content, restraining government expenditures could improve the current account directly.

15/ Devaluation is thought to affect domestic inflation through two channels. The first is the direct effect through the won price of imports since imports are important intermediate inputs for many of the products that enter the won price index for Korean manufacturers. The second channel operates through the curb market interest rate. Devaluation raises this interest rate and thereby raises costs of production for Korea's highly leveraged manufacturing sector. These costs can be passed along in the form of higher prices. Empirical verification of the inflationary effects of devaluation in the Korean context is provided in Kwack (1985) and of the transmission channels in Van Wijnbergen (1981).

Box 2.3: THE DETERMINANTS OF THE REAL EFFECTIVE EXCHANGE RATE

The effective value of a county's exchange rate depends on a set of real and monetary variables which reflect the strength of the demand for (and supply of) the country's currency. Among such variables could be included the terms of trade, the amount of capital inflow, the rate of productivity growth, and measures of domestic macroeconomic policy such as the level of public expenditures, the growth of domestic credit and the level of the nominal exchange rate.

The regression analysis reported here takes the logarithm of the REER index to be the dependent variable and suitably modified measures of the above variables as independent determinants of this index. The terms of trade variable (T) measures the relative price of imports to exports. This variable is affected both by nominal depreciation and by external price developments. The capital flow variable used (CINF) measures the ratio of capital inflows to GNP. It is expected that the higher this ratio the greater the tendency for the exchange rate to appreciate. The rate of productivity growth is most appropriately measured as a differential between Korea and its major trade partners since it is the differential that determines the net demand for individual currencies involved. It is expected that the higher the differential in favor of Korea the greater the tendency for its exchange rate to appreciate. In the empirical analysis such differentials are proxied by the average output growth differential between Korea and its major trading partners (PGD). The effects of domestic macroeconomic policy are captured through three variables; the ratio of public expenditures to GNP (PER); the rate of growth of domestic credit (GDC); and dummy variables (D and D(-1)) denoting years in which nominal devaluations occurred.

Estimation of the above model, by the method of ordinary least squares corrected by first-order serial correlation and using annual data from 1964 to 1980, yielded the following results (- denotes appreciation):

$$\log REER = \underset{(10.08)}{0.55} - \underset{(-4.6)}{0.58}*\log T - \underset{(2.5)}{0.36}*\log (1 + GDC) + \underset{(0.23)}{0.15}*\log (1 + CINF)$$

$$- \underset{(-5.4)}{0.72}*\log PER - \underset{(-1.4)}{0.78}*\log (1 + PGD) + \underset{(1.4)}{0.53}*D + \underset{(3.4)}{0.98}*D(-1)$$

These results show that all coefficients with the exception of the capital inflows have the expected signs and that most of them are significant at conventional levels. Particularly important are the signs and magnitudes of the domestic credit growth and public expenditure ratio coefficients. They indicate that in the past expansive monetary and fiscal policies have been associated with appreciating real effective exchange rates. This means that the authorities need to be particularly careful to avoid expansionary macro policies in order not to undermine a nominal devaluation. These results also indicate--through the value of the coefficients of the devaluation dummy variables--that nominal devaluations have, with with other things given, a positive effect in the short run on the real exchange rate. That means that by managing the nominal exchange rate--while maintaining an unchanged monetary and fiscal policy--it may be possible to manipulate the real exchange rate in the short run. For details see Edwards (1986).

decline in energy or oil dependence in particular since the late 1970s. Once again it appears that favorable external as well as supply side developments have worked to offset the potentially adverse consequences of macroeconomic policy.

2.42 Finally, arranging the devaluation in small and gradual steps has moderated debt service difficulties for the highly leveraged corporate sector. The increase in exports and profits has apparently kept pace with the increase in the debt service that the corporate sector has had to bear because of the devaluation. As a consequence, the policy of devaluation has not attracted much opposition from business circles.

E. Lessons for Future Policy Making

2.43 Korea has approached its stabilization and adjustment tasks creatively and in-so-doing it has benefitted from not relying exclusively on one approach or one policy instrument. It has also benefitted in that it has not applied any one instrument with great force and thereby not risked a sharper, more adverse policy tradeoff in the form of a temporary, but disruptive, recession. From the demand side it attacked inflation and an unsustainable BOP deficit by means of restrictive monetary and fiscal policies. The possible adverse effects of contractionary demand management were offset by supply-side policies which sought to raise efficiency and productivity. Incentives to promote savings and investment and policies to reduce distortions in the trade and financial regimes have helped maintain output even as the economy was otherwise being cooled down and a revolution of falling expectations being brought about. Accentuation and attenuation of the demand-side and supply-side effects has been judiciously achieved, as and when necessary, by a flexible exchange rate policy.

2.45 The consequences of macroeconomic policies may differ in the future to the extent that certain key macroeconomic relationships appear to be changing. For example, structural changes appear to have been brought about by recent changes in what may be called the "rules of the (economic) game:" specifically, the Government's progressive abstention from intervention in the economy appears to be changing the investment and savings behavior of the private sector. The implicit socialization of business risk that was the hallmark of industrial policy and development in the 1960s and 1970s is no longer to be counted upon now that Government has announced an industrial policy which limits active industry-specific intervention and emphasizes functional incentives. When fully effected, this transformation should make private investment more sensitive to monetary policy exercised through interest rates and to fiscal policy exercised through variations in tax rates.

2.46 Similarly, the mobilization of private savings has been enhanced by recent financial policy which has featured rising interest rates (in the primary banking system), a proliferation of savings instruments, and declining inflation. Should this policy persevere, additional gains could be reaped in the form of increases in financial and investment efficiency. However, opposition to further financial liberalization could become stronger if the external environment is not as supportive of growth in the future. A foretaste of this came in 1985 when deteriorating exports and profits fed industry opposition to Governments financial policy and succeeded in getting credit controls (on conglomerates) lifted and nominal interest rates rolled back slightly.

2.47 Structural changes have also been wrought by the sustained experience of stabilization and adjustment in recent years: specifically, both inflationary expectations and energy and import dependence appear to have been reduced. These developments would appear to increase the latitude for expansionary macroeconomic policy. The risk of sizable increases in inflation or the BOP deficit from expansionary policy is less now than was the case in the early 1980s. However, this must be set against the possibility that administrative wage and price guidance may also have played a considerable role, for example, in the reduction of inflation. If such repression has indeed been important in recent years, it is possible that there may be greater resistance to wage and price austerity in the future. This would impart greater rigidity to the aggregate supply responsiveness of the economy. Another development that would have a similar effect is a reduction in the growth of productivity. If, as is possible, recent productivity advances have occurred largely because of increases in capacity utilization, then the rate of such advances in the future will diminish as fuller rates of utilization are achieved.

2.48 The effects of changes in the external environment have been described at several points in the foregoing. It is clear that growth of GNP and imports in the OECD countries exercises a powerful influence on Korea's macroeconomic performance and on the effectiveness of Korea's macroeconomic policies. Korea remains an open economy and while some structural developments (e.g., successful import substitution and generalized reduction of import dependence) may make it less vulnerable to external shocks in the future, Korea is a long way from acquiring the sort of independence that Japan and the US enjoy on account of their large domestic markets. Furthermore, the high debt that Korea has acquired in the process of growth is best managed in the future by maintaining an export orientation. Flexible macroeconomic policy, and especially a competitiveness-oriented exchange rate policy, can be of great help in negotiating external shocks of a cyclical nature. Where the shocks are structural in nature, an appropriate industrial policy is called for.

CHAPTER 3: ECONOMIC LIBERALIZATION AND THE SEQUENCING OF REFORM

A. Conceptual Issues

3.01 It has long been argued that developing countries should reduce their restrictions to international trade and become more integrated with the rest of the world. The benefits of free trade are well established theoretically, and have become increasingly well documented empirically. Trading at world prices allows a country to use its resources efficiently and permits its consumers to acquire a larger quantity of goods for a given level of income. Cross country empirical studies have documented in a very convincing way that countries with more open economies have outperformed, in terms of growth and other indicators, those nations that have relied on heavy restrictions to international transactions.[1]

3.02 The general thrust of policy recommendations favoring a more liberalized economy is not restricted to the elimination of distortions in the external sector, however. It is widely accepted that the reduction (elimination) of distortions in all sectors would be desirable and for this reason, traditional policy prescriptions call for eliminating price controls, allowing (real) interest rates to rise, reducing import tariffs and quotas, and increasing the degree of integration of the economy with international financial markets.[2]

3.03 Although there is widespread agreement on the desirability of pursuing these reforms, the problem of recommending a specific sequencing for these policies is difficult on both theoretical and practical levels. Many times, due to political or other constraints, it is not possible to pursue the liberalization of all sectors simultaneously. In those cases, the question of sequencing becomes very important. Until recently, very little analytical work had focussed on issues such as: should domestic interest rates be raised before, after or at the same time as capital controls are lifted? Should the trade account be opened up before the capital account or vice versa?[3]

1/ See Little, Scitovsky and Scott (1970), Krueger (1978), and Bhagwati and Srinivasan (1979).

2/ From a theoretical perspective there is a question related to the benefits of partial reforms where only some distortions are eliminated. This, of course, is a typical second best problem. See the discussion in Krueger (1985). Of course in a textbook world with no distortions the question of appropriate sequencing is trivial. All markets should be liberalized simultaneously and instantaneously. In the real world, however, complications arise.

3/ See, however, the studies by Edwards (1984), and Krueger (1985).

3.04 Given the present commitment of the Korean Government to liberalize the economy, by dismantling trade controls and raising interest rates, the question of sequencing becomes current and increasingly important. In this section some of the more important aspects of the sequencing of economic reform will be analyzed. The discussion will basically deal with the order of liberalization of the current and capital accounts, and will emphasize macroeconomic effects of alternative sequences. The analysis will also deal with efficiency and credibility issues. The problem of determining the appropriate sequencing of reform is particularly difficult from an analytical point of view inasmuch as the analysis depends heavily on the initial conditions. In Korea, these conditions are quite different from those encountered in the most publicized liberalization attempts in the Southern Cone of Latin America, and thus, the results emerging from that latter experience may not directly apply to the Korean case. Still, the Latin American experience offers valuable lessons.[4/]

3.05 How might conditions in Korea in 1986 be characterized as it approaches further liberalization? Very simply put, these conditions are: (i) a negligible inflation; (ii) basically balanced public sector finances; (iii) a very open economy with some export bias; (iv) a repressed and segmented capital market, where an official market with fixed rates coexists with a spectrum of less controlled markets, including a freely functioning unorganized (curb) market; (v) the existence of import restrictions which are being dismantled; and (vi) selective controls to capital movements, based on a "negative list," where inflows of capital have to be approved on an individual transactions basis.

3.06 One aspect of sequencing that is well established is that impediments to capital movements (both inflows and outflows) should not be fully relaxed before liberalizing the domestic financial sector. The rationale is that if the capital account is liberalized at a time when domestic interest rates are fixed at arbitrarily low or negative real levels, an outflow of capital will result.[5/] It is clear that Korea has begun to take some of the steps necessary to bolster the domestic financial systems in anticipation of ultimately opening its capital market, but considerable

---

4/ Much of the discussion on the sequencing of reform has been done using Latin America as the benchmark. See McKinnon (1982), Edwards (1984, 1985), Corbo and de Melo (1985), and Edwards and van Wijnbergen (1986).

5/ Capital flight has indeed been the result of acute imbalances between domestic and international real interest rates of a number of developing countries, i.e., if the domestic interest rate is below the world interest rate, appropriately corrected for expected devaluation and other risk premia, financial capital will tend to exit. Much of the capital flight that Argentina experienced in the early 1980s, for example, was in response to the fact that the (anticipated) interest rate was perceived to be well below that in the rest of the world. In the current Korean case real interest rates are positive--although probably below their equilibrium level--and it is not likely that a relaxation of exchange controls would result in massive capital flight.

further reform will be needed to make this possible. Positive real interest rates now exist, but well functioning capital markets, ready to compete with external markets, would also require less segmentation, a wider range of maturities, and some financial instruments to hedge exchange rate risk effectively. Thus, in some sense, it is possible to say that Korea has taken the initial steps towards financial liberalization.

B. The Capital Account, the Current Account and the Real Exchange Rate

3.07 Perhaps one of the most critical issues related to the dynamics of economic reform is the sequencing of liberalization of the current and capital accounts. Different sequences of the liberalization will imply different paths for the critical macroeconomic variables, in particular the real exchange rate. In general, the opening of the capital account and relaxation of exchange controls in a developing country will generate important movements in financial capital. In most countries, if the fiscal deficit is under control and the domestic financial market liberalized, the opening of the capital account will very likely result in significant inflows of foreign capital, triggered by perceived differentials between the domestic and foreign returns to capital.[6/] These additional funds from abroad will allow domestic economic agents--households, firms, and the Government--to increase their expenditure over preliberalization levels. This additional expenditure will generally fall both on domestic and international goods. The increased expenditure on international goods will generate the current account deficit that usually mirrors the capital account surplus generated by the increased capital inflows. On the domestic side, the increase in expenditure will be translated in a higher demand for domestic (or nontraded) goods, and in a higher relative price for these goods. This relative price increase or real exchange rate appreciation is one of the most important consequences of a liberalization of the capital account in a capital importing country.

3.08 Under these conditions, the opening of the capital account will generate a real appreciation; however, a successful liberalization of the trade account (i.e., reduction of import tariffs and elimination of import quotas) will generally require a real depreciation of the domestic currency. This real depreciation will help the exportables sector expand, as the new structure of relative prices replaces the old protective structure, and consequently will be instrumental for reestablishing external equilibrium after import restrictions have been reduced. If, however, the opening of the capital account precludes this real depreciation, then the transition in the exportable and importable goods sector from a protective to a freer environment will become more difficult. The appreciation generated by the relaxation of capital controls will squeeze profitability in the tradable goods sector precisely at the moment when it is going through a costly readjustment. This would be a major risk for Korea in prematurely opening up its capital market.

6/ This assumes that the developing country has relative abundance of labor, and scarcity of capital. In the present world economic conditions with many LDCs affected by the international debt crisis, an opening of the capital account may not result in massive capital inflows.

3.09 Consequently, it has been suggested that the capital and current accounts should not be opened simultaneously and, moreover, that during the transition period after trade has been liberalized, capital inflows be carefully controlled.[7/] This view is reinforced by the fact that, in general, the opening of the capital account will result in an overshooting of capital inflows, which will provoke a steep real appreciation in the short run. In other words, experience has shown that immediately after the relaxation of controls, capital inflows are likely to be quite large, but that after some time they decline towards a new long-run equilibrium level. There are a number of examples where the opening of the capital account has resulted in massive capital inflows and a significant real appreciation of the domestic currency.[8/]

3.10 The conflicting movements of the real exchange rate as a result of opening the capital and current accounts (i.e., real appreciation and depreciation respectively) capture the fact that these policies will exercise pressures for resources to move in opposite directions. To the extent that there are adjustment costs associated with resource movements between sectors it is advisable to implement policies that would avoid unnecessary resource switches, i.e., resource movements that will be reversed after a short period of time.[9/]

7/ See McKinnon (1973, 1982), Dornbusch (1983), Frenkel (1982), Edwards (1983, 1984).

8/ Perhaps the best known recent case was Chile during 1979-81 when massive capital inflows resulted in a reach exchange rate appreciation of 30%. The Korean experience during the 1960s has also been cited in connection with capital inflows following a capital account liberalization. In mid-1966, a large inflow of short-term financial capital flooded the Korean economy. As a result, there was an increase in inflation and a substantial real appreciation. It should be noted, however, that throughout their experiences with major inflows of capital, Korea and Chile followed different strategies. Whereas in Korea the nominal exchange rate was adjusted periodically, in Chile a fixed nominal rate was in effect during most of the period. In both countries, however, a real appreciation took place, with the real exchange rate moving against exporters.

9/ Once more Chile provides a good example. Starting in 1976--after the path of trade liberalization had been announced--exporters embarked on extensive investment programs aimed at increasing their export capacity, only to find that in 1979-81 the real exchange rate turned drastically against them. At this point investors moved away from the export sector and switched their resources to the nontradable sectors and especially towards construction. It has been suggested that one simple way of avoiding these unnecessary resource switches is by opening the current account first, and, only after the new productive structure has been well established, opening the capital account slowly. See McKinnon (1982), Frenkel (1983).

3.11 The preceeding discussion has assumed that once the capital account is opened, domestic agents would be able to borrow from abroad and capital would flow into the country. This is indeed what would be expected under normal conditions, where uncontrolled (real) domestic interest rates would be substantially higher than abroad. This would, in fact, very likely be the situation in Korea if all exchange controls were suddenly lifted. However, given the current mood in the international financial community regarding lending to the developing countries, there is a risk that these incoming funds might be of short-term maturities, so that the real appreciation which would occur might be of short duration.[10/]

C. Economic Reform, Efficiency and Credibility

3.12 Policymakers considering the macroeconomic consequences of alternative sequences of reform may wish to consider two additional issues. The first of these deals with the effects of the reforms on the efficiency of the resource allocation and consumption processes. The specific question is how the economy's overall efficiency will be affected by removing distortions in only one sector.[11/] Even though the evidence is not conclusive, there is a strong presumption that the relaxation of capital controls in the presence of tariffs will amplify existing distortions, reducing the efficiency of the economy. Once capital controls are lifted a proportion of the funds obtained from abroad will be used to increase investment in the import substitution sector. However, since tariffs have not been lowered yet, this sector is already producing "too much" and "too inefficiently" relative to what it would produce under a neutral trade regime; the existing distortion may, thus, be amplified. On the other hand, the presumption is that the reduction of tariffs in the presence of capital controls will generally not produce negative amplifications of existing distortions. On the contrary, it is

---

10/ In the current debt environment countries may face a situation of international credit rationing. Under these circumstances it is likely that the opening of the capital account will not result in additional capital inflows. Moreover, depending on the approach the country is taking to solving its debt problems, capital may even tend to flow out when capital restrictions are relaxed because in some instances the stabilization program is carried out timidly. These slow adjustments introduce significant uncertainties regarding the future behavior of the exchange rate and domestic interest rates. These considerations again point towards delaying the opening of the capital account until the smooth functioning of domestic capital markets has been demonstrated.

11/ According to the "second-best" theorem of welfare economics, if existing restrictions are only relaxed sequentially it is not possible to know a priori whether some of the remaining distortions will be magnified due to the partial liberalization. If this is the case, partial reforms may generate negative results and if the magnification of the remaining distortion is big enough, partial reforms may have a negative overall short-run effect on the economy.

likely that a positive indirect effect will result because the reduction in tariffs will likely result in a higher demand for foreign funds.[12/]

3.13 The second important issue is related to the degree of credibility of economic reforms. Credibility will generally affect the perceived path of relative prices and incentives. If a trade reform announcement is credible, firms and investors will expect a particular set of future movements in prices and relative returns to investment to ensue and they will react accordingly, mobilizing resources, and investing in the "new export industries." On the other hand, if the public believes that the reform is not credible, and that there is some probability that the reform will be reversed in the future, "cheap" foreign funds, obtained through the opening of the capital account, may be used by the owners of firms in the import substitution industries to maintain their firms functioning at a (temporary) loss until such time as the reforms are reversed.

3.14 It is important to stress that the degree of credibility--which is critical for the analysis of the sequencing of liberalization--should not be viewed as completely exogenous. On the contrary, the strategy followed during the liberalization process will tend to affect this credibility. A fundamental and critical aspect of establishing credibility is related to the internal consistency of the policies that have been and are currently being pursued.[13/] This aspect of reform has been a strong point of the Korean efforts inasmuch as Government has used a long preannouncement period for basic reforms and has stuck to it. Credibility has thus not been a problem in the Korean context.

3.15 Summarizing, although there is no ironclad rule regarding the appropriate sequencing of liberalization, both analytical considerations and historical evidence suggest that a prudent policy consists of liberalizing the current account well before the capital account is fully opened. A corollary is that the domestic financial sector must be sufficiently strong and the capital market sufficiently developed to allow the benefits of capital account liberalization to be reaped. Premature liberalization, before domestic institutions are capable of operating effectively is usually ill-advised. Equally important perhaps to the sequencing of reform is the confidence these policies inspire. In that sense, a crucial aspect of the liberalization process is to define a consistent and credible policy package.

12/ See Krueger (1985) and Edwards and van Wijnbergen (1986).

13/ For example, the inconsistency between fiscal and exchange rate policies in Argentina in the late 1970s, played an important role in the eventual abortion of that country's liberalization program. Also, in Chile, many agents thought that since tariff reduction was accompanied by a significant real appreciation of the exchange rate, the trade reform was actually unsustainable. The extraordinarily large current account deficit observed in that country during 1979-81 further fueled the feeling of the reforms being unsustainable. This aspect of credibility should be an important consideration for policymakers undertaking trade policy reforms.

D. Review of Sequencing and Implementation

3.16 The Korean Government's commitment to liberalizing the economy is clearly serious and reforms are progressing at a steady pace. The general approach followed has been gradual and prudent. Progress, in terms of liberalization, has been different across sectors, with reform of the trade account proceeding at a relatively more rapid pace than the reform of the internal domestic market, and of the capital account.[14/]

3.17 The fact that the liberalization of trade--movement of items to the AA list and reduction of tariffs--has proceeded in a preannounced, gradual fashion is a clearly positive step. This strategy, if well designed, will help those firms affected by the reform to adjust in a smoother fashion. Generally, the problem with this type of policy is to define a time span for the reform to be carried out that is sufficiently long, so that the adjustment can be indeed smoother, and sufficiently short, so that the announced liberalization measures are credible. To date it seems that policymakers have been successful in defining this timing. The reforms are credible--although interest groups still lobby against it--and those firms that expect to be affected by it are taking some measures towards adjustment. On the other hand, until now no major disruptions in production associated with the reforms have been observed, which can either imply that initial trade restrictions were not terribly binding, that they still are operative in some guise, or that recent real depreciations in the exchange rate have had their desired effect. As noted earlier, changes in the degree of coverage of the AA list provides biased information, since many items are still affected by special laws. This procedure reduces the transparency of the importation process and it is recommended that the necessity of these special restrictive laws be reviewed. If necessary for their abolition, a temporary hike in their respective tariff levels could be considered, and would probably be preferable to any binding constraints imposed by these devices.

3.18 It is important to realize that the purpose of a trade liberalization reform is to increase the degree of openness of the economy and to improve the domestic sector's efficiency, The purpose of these reforms is not to generate a large trade deficit. This means that, in order to increase the amount of imports without increasing the trade deficit, exports should increase pari passu with imports. This has to be achieved through higher incentives to exports. These incentives, in turn, come from two sources. Generally by liberalizing imports, imported intermediate inputs are made available at a lower price to the exports industries. Second, in order to maintain trade equilibrium, liberalization must usually be accompanied of a real depreciation. In the current Korean experience this real depreciation

---

14/ Liberalizing the trade account before the capital account and the domestic capital market account before the external capital account are both clearly warranted, although there is no clear consensus concerning the appropriate sequencing between trade and domestic finance. See Park (1985).

has indeed taken place; in 1985 alone the real effective rate fell by 12%. In that regard the Korean reform is proceeding in the way most experts would recommend.[15] It should be emphasized that in the next few years the evolution of the real exchange rate will continue to be very critical for the behavior of the external sector.

3.19 The strategy followed in Korea has also been prudent regarding the sequence of liberalization of the current and capital accounts. As discussed, there are strong presumptions that the liberalization of the capital account should be postponed for a substantial period until after trade reforms have been finalized. By maintaining some control over capital movements, erratic and counterproductive changes in the real exchange rate can be avoided. The current policy of monitoring capital movements is therefore quite important during the next few years to ensure that there is little impetus for backsliding on trade reform.

3.20 An important aspect of the Korean reform which has not generated sufficient attention is related to the liberalization of the service sector. Korea is preparing to open selectively its insurance market to foreign underwriters in 1987, for example. Some care should be taken to avoid confusing the liberalization of the service sector, which has benefits if the cost of those services to Korean industry can be reduced, with a selective opening of the capital account. For example, this could be the case, to some extent, if foreign insurance companies operating in Korea were allowed to freely diversify their portfolios internationally, with resulting surges in capital outflows.

3.21 Overall the liberalization of Korea's external sector is proceeding smoothly and deliberately. It is noteworthy that liberalization proceeded according to schedule, despite dim export performance in 1985 and ensuing lower growth. This reflects a genuine commitment to liberalization. At the same time, restraint is being exercised during the transition. In light of the still somewhat immature state of domestic financial markets, and the fragility of certain aspects of corporate finance, this is prudent. The next priority in terms of achieving overall liberalization objectives should be the bolstering of the financial sector and, in particular, financial institutions. In order to compete effectively with foreign firms in financial markets, greater attention to profitability and risk will be needed, and some of the kinks endemic in the Korean system of industrial finance will need to be ironed out. Premature opening of the capital account is to be avoided until such time as domestic capital markets achieve the requisite level of stability and sophistication.

15/ While it is true that some of the real depreciation has been quite unintentional, and has basically been the result of the Yen's appreciation, Korean policymakers opted to accept the lower value of the won.

FINANCIAL LIBERALIZATION RISKS: LESSONS FROM THE SOUTHERN CONE

A3.01 The recent Latin American Southern Cone experience with liberalization has been traumatic, and has had an important effect on conventional thinking. Interventionists have found in these failed attempts new ammunition with which to attack the merits of market oriented development strategies; supporters of the liberalization process, on the other hand, have been forced to rethink their views on the optimal path to be followed during the transition period of a liberalization attempt. The Southern Cone experiments provide very useful lessons that should be closely scrutinized by policymakers who want to avoid the mistakes made, especially during the transitional period, which eventually contributed heavily to the collapse of the reform experiments.[1/] In this section some of the more relevant aspects of the Southern Cone experience will be analyzed, and the potential risks for Korea will be highlighted. Most of the analysis will focus on the case of Chile, since it was in this country where the reforms were pushed furthest.

A3.02 It should be stated at the outset that the current economic conditions in Korea are significantly different from the initial conditions encountered in the Southern Cone in the mid-1970s. First, Korea does not have a serious inflation problem, while rampant inflation was perhaps the most urgent problem in Argentina, Chile and Uruguay. Second, the fiscal sector in Korea is under control; in the Southern Cone, on the contrary, the fiscal deficit was initially gigantic, and in Argentina it was never really reduced to manageable levels. Third, Korea's external sector is very open, and has had a long successful tradition of export growth. Fourth, Korea's financial sector, while controlled and therefore distorted, has been in place for a long time, and has in the last few years been operating with (significantly) positive real interest rates. In the Southern Cone--and particularly in Chile--on the contrary, the financial sector was in shambles when the reforms were first started. All these factors make the cases of Korea and the Southern Cone quite different and comparisons should be handled with caution.[2/]

A3.03 Perhaps the two more striking features of the financial reforms in these Latin American countries were the high real interest rates, and, particularly in Chile, the large proportion of nonperforming loans that the banks accumulated. Immediately following the liberalization of interest rates, these became extremely high in real terms. For example, in Chile they averaged more than 40% for long periods of time. At the same time the maturity of bank operations was very short, seldom exceeding 30 days. These

---

1/ See Bruno (1985), Corbo (1985), Corbo and de Melo (1986), and Edwards (1985).

2/ There are, however, perhaps surprisingly, some similarities between Korea's and Chile's financial reforms. In both cases, big conglomerates bought major interests in the national banks. Also, in both cases banks had a large proportion of nonperforming loans, firms were very highly leveraged and reportedly could actively use borrowed funds as equity capital.

high rates persisted throughout the period, in spite of the massive inflow of capital that followed the relaxation of exchange controls. Most domestic banks in Chile, and to some extent in Korea as well, were basically owned by large conglomerates, accumulated a large proportion of nonperforming loans.[3/] At the end, in Chile--and to a lesser extent in Argentina--the combination of high real rates, large proportion of bad loans in the banking sector, and widespread expectations of devaluation conspired to generate a major collapse of the financial sector. In Chile several conglomerates and banks failed, and had to be rescued by the Government. In Argentina a number of banks also had to be bailed out by the Government.

A3.04 With hindsight it is possible to say that perhaps the most serious problem related to the reforms of the domestic capital market, both in Chile and Argentina, was the lack of effective supervision of the financial sector. For example, in Chile, as early as 1974 it was well known by everyone involved--including, of course, the "regulators"--that the different conglomerates were finding ways to circumvent the rules covering excessive ownership concentration in the financial sector. In fact, it was known that these conglomerates had managed to control most of the largest banks.[4/] Starting in 1974-75, many banks concentrated large fractions of their loan portfolios on related firms owned by the conglomerate that controlled that particular bank. The basic scheme used by many of these conglomerates was to use the financial resources obtained through the newly acquired banks to buy firms that were being privatized; some of these funds were also used to expand the level of operations of these and other firms. However, many of these loans to related firms did not represent, from a purely financial point of view, sound banking practices. In fact, banks were used as pawns in the conglomerates strategy of "growth at almost any cost" in a tightly speculative environment.

A3.05 It has been argued that the existence of an evergrowing "false demand" for credit by the conglomerates played a central role in the explanation of the high interest rates.[5/] This "false demand" consisted of the rolling-over of loans, which in turn, had their origin in the privatization of a large number of firms during the early years of the military regime. As pointed out above, many of these firms had to spend significant resources in order to operate, modernize and expand these companies. Many of them did not turn out to be profitable, and increasingly resorted to additional borrowing in order to stay afloat.

A3.06 The financial problems faced by banks and firms alike were compounded, both in Chile and Argentina, by a steep real appreciation of the domestic currency. This tendency towards overvaluation was the result, among other things, of the exchange rate policies followed in these countries which consisted of preannouncing a rate of devaluation significantly lower than the ongoing rate of inflation--and of the massive inflows of capital that followed

---

3/ See Table 5.16, Volume I.

4/ Another major difference is the vigilant role played by the Ministry of Finance in Korea, and the close coordination among economic ministries.

5/ Harberger (1985).

the opening of the capital account of the balance of payments.[6/] One of the most important consequences of the real appreciation of the domestic currencies was that the tradable goods sector suffered an important loss of competitiveness, and that a large number of firms ran into serious financial trouble. In both Chile and Argentina most of these firms faced these problems by resorting to heavy borrowing from the financial sector, exercising additional pressure on the demand for credit and on interest rates. This circle of events came to a sudden end in both countries with major exchange rate and financial crises.

A3.07 The main lessons--in terms of reform of the financial sector--that emerge from the Southern Cone fall into four categories. The first category refers to bank supervision, It is essential that the process of financial reform is accompanied by strict supervision of the banking and financial sectors, in order to avoid fraud, circumvention of sound banking practices, or misuse of the financial resources. This supervision should be particularly strict when conglomerates play such a crucial role in the structure of industrial organization, as in Chile and Korea. Strong supervision should reduce the chances of a major bank collapsing with the attendant dangers for foreign borrowing, in which creditworthiness is dominated by a herd mentality among lenders.

A3.08 The second area where lessons emerge refers to the speed and extent to which interest rates should be freed. The Southern Cone case suggests that a prudent attitude may be warranted when dealing with financial liberalization. More concretely this means that interest rates should be raised slowly, with an effort being made for maturities to be extended. The slow and prudent raising of interest rates will allow the authorities to detect and tackle early on any market reaction based on destabilizing expectations or lack of credibility. A precipitious rise in the cost of borrowing would merely serve in a highly leveraged corporate environment to distress financing and greater instability.

A3.09 The third lesson relates to the liberalization of the capital account and the relaxation of exchange controls. A number of authors analyzing the Southern Cone experience have concluded that the timing used to open the capital account was ill-conceived in all three countries.[7/] As argued in Section I of this chapter there are compelling reasons to suggest that the opening of the capital account should be pushed significantly back until after the trade account has been opened. The fourth lesson relates to exchange rate policy. It is extremely dangerous to reform the financial sector in an environment of overvalued domestic currency. If this happens, as in Chile and Argentina, large distortions in terms of "excessive" borrowing from abroad may take place, in anticipation of real devaluations. In this context, Korea's realistic exchange rate policy serves as a useful lesson of the liberalization experience.

6/ See Corbo (1985) and Edwards (1985).

7/ See Edwards (1983, 1984), Dornbusch (1983), McKinnon (1982), and Bruno (1985).

CHAPTER 4: THE THEORY OF INDUSTRIAL POLICY IN THE KOREAN CONTEXT

A. Incentive Regime

4.01 Many arguments are offered to justify industrial interventions. On the basis of economic criteria, there are only a few clearly delineated cases for such public actions. In general, intervention can be justified by evidence of nonexistent or malfunctioning markets.[1] In developing countries, imperfections are most often claimed to occur in markets for capital, technology, and knowledge. But often far more important is the economic environment in which these markets operate, namely, the incentive regime.

4.02 The general structure of incentives is often the most important element of industrial policy in developing countries. The consensus among development theorists favors an outward-oriented incentive regime with relatively uniform incentives for different production activities. The case for such a regime rests, in the first place, on the proposition that a country can maximize its income and growth (at world prices) by shifting resources into areas of comparative advantage and later by adjusting this allocation to track changes in comparative advantage. An outward-oriented regime also reduces uncertainty about the government's priorities and policy directions. The outward-oriented strategy has received strong empirical support from evidence accumulated in postwar development history.[2]

4.03 The benefits of outward orientation appear to be greater, in fact, than what might be reasonably attributed to the achievement of allocative efficiency alone.[3] Korea's openness has provided positive externalities in

---

1/ Even this evidence is theoretically insufficient, to warrant intervention, however. Since a compensating policy may cause undersired side-effects, there must be also a strong presumption that the economy will be closer to its "first-best" optimum with intervention than without.

2/ Work by Balassa, Krueger and others has shown that countries with outward oriented regimes grew faster than those with strong incentives favoring production for domestic markets. Improved performance was also commonly observed for countries which changed to an outward oriented regime during the postwar period. Needless to say, Korean experience figures importantly in this research. See, for example, Frank, Kim and Westphal (1975).

3/ Krueger (1985) and others argue that open economies are more flexible than import-substituting economies because they have greater "cushions" of unessential imports to squeeze under adverse circumstances. According to Westphal, Kim, Dahlman (1984) and others, openness can also facilitate the absorption and mastery of foreign technologies. Nishimizu and Robinson (1983) have shown that the rate of technical progress was faster in countries with outward oriented policies than in those pursuing an import substituting strategy.

the areas of macroeconomic flexibility and the absorption of technical progress. These externalities justify even greater participation in trade than would be optimal from the allocational viewpoint alone and validate Korea's past proexport bias. While recent analyses agree that an outward orientation is conducive to rapid development, the precise implementation of such a policy is somewhat more controversial. According to some, it implies low, uniform levels of effective protection, with perhaps some escalation to permit infant industry development in higher stages of assembly and production. Others see a somewhat greater role for selective support for outward oriented infant industries. Despite these differences, however, there is wide concensus that protection ought to be lower and more uniform than is currently the case in Korea as well as in most other developing economies. It is in this light that the Korean trade liberalization, described in Chapter 2 of Volume I, is quite important, as a natural successor to the effective export-promotion activities of the past two decades.

B. Functional Intervention

4.04 Since the case for the open economy assumes well-behaved markets (for factors of production as well as final products), the absence or imperfection of markets provides a key rationale for nondiscriminatory functional intervention. This rationale points to policies that correct the market failures themselves or compensate for them throughout the economy by offering functional incentives. It does not provide an argument for selective intervention in favor of specific sectors or subsectors, however. To be sure, the implementation of corrective measures may involve a sectoral focus and sectoral coordination, but efforts to correct market failures should not generally require comprehensive programs of aid to a specific industry.

Technology Development

4.05 The development of technology generates positive externalities if the benefits of innovation "leak out" from the innovating firm to its competitors. In the absence of compensating public subsidy, firms will invest less in research than is justified by social returns. The decision of individual firms to allocate revenue to R&D expenditures may be influenced also by their competitive setting and by the minimum scale required to undertake specific forms of R&D activities. There may, therefore, be an added role for Government in subsidizing or coordinating research where competitors cannot effectively pool their efforts for market reasons. In cases where R&D investments are socially suboptimal, tax incentives can be used to increase the implicit private rates of return on investment, while in cases where private scale is inadequate, coordinated (public or private) activities are legitimate alternatives.

4.06 Not all research externalities justify public intervention, however. To merit public support, a research project should generate externalities for other national firms, but not for firms in other countries. While the general case for public policy toward research is widely recognized, few governments appear to apply the test of "commercial externality" to guide the specifics of their technology policy. In the Korean high technology sector, for example, individual firms are relatively large, while the sector as a whole is relatively small (as compared, for example, to its OECD competitors). Thus implies that external benefits from the research of a large Korean firm are relatively more likely to accrue to foreign competitors and relatively less likely to accrue to other domestic firms.

4.07 Intellectual property rights--protection afforded to inventors through patents and copyrights--represent an important, related area of policy. The problem is familiar; the strict protection of property rights raises the price of new products above marginal cost and (from a static economic perspective) excludes too many potential users. On the other hand, it provides strong incentives for innovation. The optimal system for a given country depends on the net balance of innovations produced and used, and on the effect of property rights laws on the inflow of foreign technology.

Manpower Development

4.08 Education and training are hard to finance privately because of the absence of institutions (families excepted) that lend to individuals against uncertain and distant future income returns. Thus, market forces generally lead to underinvestment in human capital. In addition, advanced training and foreign training may involve externalities: the benefits of inventions and technological transfers made possible by technical and foreign training may not wholly accrue to any one employer. Consequently, wages will fail to reflect the contribution of advanced training. All these possibilities argue for some public subsidy.

4.09 Retraining can be justified by the same arguments as education in general, but additional motives may be present. Retraining provides targeted assistance to people who are suddenly and adversely affected by changes in trade or production patterns. In this context, retraining helps to achieve distributional goals as well as increasing the political acceptance of changes in industrial structure. Moreover, retraining increases the flow of labor from declining sectors to growing sectors. Thus, retraining constitutes a "second best" policy where factor mobility is hampered.

Capital Markets

4.10 Underdeveloped capital markets represent a frequent rationale for government intervention in the allocation of credit. In many cases, however, the problems attributed to faulty capital markets represent not so much market imperfections as the high real costs of providing or administering credit. For example, markets may be "missing" for financing new ventures if the number of such ventures is too small to support adequate risk pooling. Similarly, domestic capital markets may be too small to diversify risks for the establishment of new large-scale industries. But the problem of small risk pools is a real problem whether credit is publicly or privately allocated; to reduce such risks it is necessary to enter arrangements that pool risks across countries. These arrangements may be either private or public. Similarly, the high cost of capital to small or new businesses often reflects the absence of organized credit information on businesses and entrepreneurs. This makes the real costs of administering credit high; it would not be efficient to increase the flow of credit to such borrowers at subsidized or controlled rates.

4.11 A more compelling rationale for public intervention in credit markets must rest on unequal access to credit for reasons other than the high real costs of credit. In the Korean case, it has been argued that borrowers that are large compared to relatively isolated domestic credit markets apply strong bargaining (monopsonistic) powers to achieve unusally good terms on loans. Large companies may be getting low cost loans also because they enjoy an implicit government guarantee--that is, they are more likely to be rescued than small firms. These examples provide a rationale for compensating government intervention. In designing an intervention program, however, it must be recognized that costs of providing credit to large firms are generally lower than for small firms, due to differences in administrative costs, information availability, and risks of bankruptcy. Thus a differential in borrowing rates in favor of large firms is not, in itself, evidence of market imperfection.

Competition Policy

4.12 Governments in the industrialized countries frequently act to preserve or enhance competition in product markets. For smaller countries, such as Korea, the need to exercise control, at least with regard to industrial subsectors, is closely linked to the extent of protection. In the case of an open economy world markets provide ample competitive pressure. In the case of protected markets and markets for nontraded goods particularly markets that are small relative to the minimum efficient scale of production, large firms may exercise market power with adverse implications for consumer welfare, innovation, and product development. The problem is especially serious if protection is administered by quantitative restrictions rather than tariffs.

C. Selective Intervention

4.13 Selective intervention, defined as policies that change the allocation of resources among specific industrial sectors or subsectors can be justifed by a variety of arguments. First, specific externalities and market failures may warrant intervention in a particular industry. Second, government intervention may be also essential (and profitable) in support of firms that are competing in internationally oligopolistic markets. Third, noneconomic objectives may exist with respect to industrial structure, such as greater economic independence, or greater capacity to produce arms or transport the nation's essential goods. Fourth, industrial policy objectives may also include other, quasi-economic aims, such as the location of industry, employment, and income in certain geographical areas or the desire to reduce the vulnerability of the economy to particular markets. It must be explicitly recognized that interventions designed to achieve these structural goals are often costly and inefficient, and that their burden must be compensated by the benefits of achieving the government's noneconomic objectives. Alternative ways of achieving industrial policy goals often differ with respect to cost as well as the distortions they creat relative to market outcomes.

4.14 There is an active ongoing debate on selective industrial policy in industrialized as well as more advanced developing countries which is relevant for Korea. In "sunrise" sectors the debate centers as why an activity ought to operate at a higher level than dictated by market forces. From an economic perspective, the case must rest either on the superior decision-making capability of government, or more likely, or interfirm externalities from learning on technical progress. In "sunset" industries, the proponents of interaction must demonstrate that the pace and/or patterns of decline dictated by the market are inappropriate, or more specifically, that market forces are leading to socially inefficient outcomes. The attendant issues which require normative judgments include (i) the desirable balance between social costs and benefits, and (ii) the extent to which the existing private institutional framework is capable of efficiently reallocating productive resources among industries.

Sunrise Industries

4.15 Two kinds of questions must be answered with respect to emerging industries: what makes an industry a potential winner and why does the private capital market fail to provide the required support? Before addressing legitimate economic arguments for selective support, it is important to note that common notions about what makes an industry "strategic" are often poorly founded. Consider, for example, the strategy of aiding "high value added" industries. The implied (and flawed) reasoning is that such industries generate high levels of national income per unit of output. Depending on how precisely "high value added" is defined, a policy that encourages such industries may lower national output, employment, and welfare (see the implications of alternative definitions in Table 4.1). At best, under the most favorable definition, the value added criterion produces the same ranking as would result from applying the conventional economic criteria of profit maximization and production according to long-run comparative advantage. In this case, other market imperfections aside, the ranking does not need to be implemented through government policy because it will simply correspond to the ranking that market agents use to decide on alternative investments.

4.16 The existence of interfirm externalities provides a more solid rationale for intervention. Such externalities may involve goods, services, and technology created by the firm (in addition to its conventional products) for which the firm is not (fully) compensated. They may include the acquisition of knowhow that is partly passed on to other firms, on-the-job training of managers and workers who later move to other businesses, or the development of markets for more differentiated and higher quality inputs, which in turn help other production activities. These externalities are variously referred to as agglomeration economies, scope economies, and linkage economies. This kind of argument forms part of the rationale for highly selective support for a few appropriate infant industries.[4/]

4.17 Assuming that intervention is justified, the size of the subsidy should just match the externality identified; it should correct the incentive system so that the production of a good is rewarded by the full social

---

4/ This corresponds to Westphal's (1982) recommendation.

benefits achieved. The rest--whether or not the industry actually develops--should be left to market forces. In practice, unfortunately, once governments commit themselves to aiding a particular industry they seldom try to provide the "correct" subsidy; rather, they adopt a package of development measures designed to make the industry grow whether it is economically viable or not. Practioners of this approach would claim that there is no such thing as a flawed industrial policy, merely an insufficiency of incentives. Put differently, there is a great temptation for policymakers not only to pick winners but also to ensure winners.

Table 4.1: EFFECTS OF TARGETING ACCORDING TO A "VALUE ADDED" CRITERION

Precise definition of criterion	Implied structural bias	Economic consequences of promoting the sectors selected by the criterion
Value added as percent of gross output	Against goods with intermediate goods requirements	Reduced productivity as structure favors sectors at the beginning of the production chain, rather than those in which value added per unit of input is highest
Domestic value added as percent of output	Against goods with high imported intermediate goods requirements	Reduced productivity as structure favors sectors that use chiefly domestic raw materials and intermediate inputs, rather than those in which economy has comparative advantage; diminished intra-industry trade and technology transfer
Valued added per worker	Against goods in which a large part of value added goes to labor, i.e., against labor-intensive goods	Artificially high demand for capital, possible unemployment, declining productivity
Value added per unit of total factor input	Against goods in which total income generated per unit of input is low	Increasing productivity; rank corresponds to economic profitability and comparative advantage (assuming no market imperfections or externalities)

4.18 A new view on sunrise industry policy, potentially relevant to Korea, is that governments can strengthen the international market power of their national firms vis-a-vis foreign oligopolistic competitors by adopting supportive policies, even including protection of domestic markets. More broadly, the realities of international trade, which include protectionist agreements as well as possible collusive behavior, may make it difficult for Government to totally distance itself from the activities of its firms. In particular, these considerations may affect government policy towards industrial organization within Korea, and towards joint ventures between Korean firms and foreign partners.

Sunset Industries

4.19 In declining industries, two quite different issues are at stake. In some cases, government policy may be needed to aid the withdrawal of resources from a declining industry. In other cases, government participation may be needed to "restructure" an industry, that is, to add resources and/or alter management in order to reach a stable and profitable operating structure. These goals involve different analytical issues. The withdrawal of resources from a declining industry does not, under usual market conditions, require government intervention.[5/] The existence of large adjustment costs, for example, is in itself not a valid reason for intervening to either slow or accelerate the market rate of adjustment. Similarly, well-functioning markets are also able, in principle, to engineer optimal restructuring. As in other dimensions of policy, intervention needs to be justified with specific market failures or imperfections, or by virtue of noneconomic objectives.

4.20 Restructuring involves turning a declining activity into a competitive one. When restructuring is in fact possible, it requires an infusion of new resources--capital, management, and technology--into an activity which is presently unprofitable. If the returns on such new investments can be clearly separated from obligations on older debt, then private capital markets can be relied on to supply restructuring capital, and in general to determine whether restructuring is feasible or not. However, the separation of "old" debt from "new" debt is not easily effected in practice. Bankruptcy is one way to do so--it wipes the slate clean so that new investors can lay claim to future returns. But bankruptcy is complicated and disruptive; it is often impractical for large economic units. In Korea, for example, the bankruptcy of a large group may create concern for the creditworthiness of the entire financial sector. In some countries, one or a few banks control a sufficiently large part of a company's debt to step in with a restructuring plan.

5/ Market imperfections may lead to inefficiently slow or fast adjustment, however; for example, high unemployment compensation in industrialized countries accelerates the decline of sunset industries (compared with ideal market conditions), as workers are reluctant to accept wage cuts that would permit their companies to compete. At the same time, unemployment compensation reduces incentives to move into new sectors (which may involve significant financial and psychic adjustment costs) and thus decelerates the growth of sunrise industries. Retraining and adjustment assistance policies represent second-best solutions for compensating for these effects.

In others, however, private capital markets could fail to support even very profitable projects in declining industries.

4.21 Public support of restructuring can take various forms. Government loans provide funds directly. Loan guarantees may assure creditors the returns that cannot be assigned to them under conventional arrangements. Cartels organized by governments may enable troubled industries to raise prices and, in effect, be rescued by consumers.[6/] Mergers among competitors may permit a rational scrapping of capacity. In all of these cases, public intervention has adverse side effects and should be pursued reluctantly; at a minimum, intervention should be contingent on a package of measures that promises a return to profitability.

4.22 It has been recently suggested that government intervention may be justified even in the absence of market imperfections in the case of oligopolistic declining industries.[7/] The argument is that oligopolistic firms sometimes continue to fight for market share even if this involves selling at below marginal cost, in the hope that they will "win" a large share of a profitable market as their competitors withdraw. If the firms of the industry are similar--equally likely to win--excessive (that is, below below-marginal cost) competition may persist for a long time, implying an inefficiently high level of output. A government restructuring plan which reduces capacity (perhaps by compensating the withdrawing firms) would improve national welfare. Though particularly appealing, this cannot justify intervention in most declining industries, such as the Korean shipping, for example, where the distribution of capacities is highly skewed, and different firms have obviously different probabilities of financial survival. Indeed, in the typical case the possibility of a government restructuring plan may in itself serve to delay adjustment, as potential losers have added incentive to wait for a government rescue.

4.23 As the Korean economy becomes more sophisticated and the cases of overt market failure shrink, the rationale for industrial intervention declines. In keeping with this trend, Government has redirected its philosophical stance in recent years to favor functional incentives rather industry-specific incentives. It has not yet withdrawn from industrial decision-making entirely, however, particularly on the side of declining industries, where public externalities include effects on employment, the capacity to borrow on international capital markets, and perhaps the integrity of the banking sector. Therefore, the emerging challenge to policymakers is to rein-in government's tentacles, while at the same time improving those markets still characterized by imperfections and bolstering those institutions which are needed to improve the functioning of those markets.

6/ This is in fact the Japanese solution. See Peck, Levin and Goto (1985) for a detailed description.

7/ Okuno and Suzumura (1985).

THE JAPANESE APPROACH TO INDUSTRIAL RESTRUCTURING

A4.01 Introduction. Recent Korean intervention in several major industries, including shipping and overseas construction, somewhat resembles Japanese policies in industrial restructuring. Since the early 1950s Japanese law has authorized the formation of so-called "depression cartels" in distressed industries, aimed at lowering output and raising prices for a limited period of time, as well as "rationalization cartels," aimed at encouraging the improvement of industry profitability through cooperation in research and product standards.[1] These laws were originally designed to deal with cyclical downturns and did not address structural adjustment problems in declining industries. As a consequence of the economic turmoil of the 1970s, a "Law of Special Measures for the Stabilization of Specific Depressed Industries" was adopted in 1978, with a five-year horizon, and it was replaced by a "Law of Special Measures for the Structural Improvement of Specific Industries" in 1983.

A4.02 Under the 1978 law an industry could be designated as a depressed industry if (a) it had severe excess capacity; (b) more than half of its firms faced "dire" financial conditions for at least three years; and (c) more than two thirds of its firms requested designation. Designation also required the approval of the affected Ministry (typically the MITI, but in the case of shipbuilding, for example, the Ministry of Transport). Designation exempted an industry from antimonopoly laws in order to permit the negotiation of a joint capacity reduction plan.[2] It also provided limited low interest loans for financing retirement allowances, and made an industry eligible for loan guarantees. In fact, the financial provisions of the Law proved unimportant; the loan guarantee fund, for example, was very much undersubscribed.

A4.03 The 1983 law has similar provisions with respect to designation and capacity reduction; however, it offers somewhat broader financial support, including (i) low-interest loans from the Japan Development Bank for modernization and improvement; (ii) grants for R&D spending, especially in the field of energy conservation; and (iii) accelerated depreciation for equipment for modernization and additional minor tax benefits. It also permits the formation of "business tie-ups" in which firms can agree to join together to fully coordinate their production and/or sales activities. These business tie-ups,

---

1/ For broader discussion of Japanese industrial policy see Magaziner and Hout (1980), Johnson (1982) and Petri (1984).

2/ The Ministry relevant to a designated industry was charged with developing a stabilization and capacity reduction plan, in close consultation with the industry's trade association and individual firms. The Government had no power to implement the capacity reduction plan, although the establishment of new plants and the expansion of existing plants was in principle prohibited. There is in fact a case of capacity expansion by a "renegade" firm under a capacity reduction program, in the electric steel furnace industry.

which must be approved by the Fair Trade Commission, resemble mergers in that the unit formed by the "tied-up" firms typically acts as a single competitor against other tie-ups.

A4.04 Operation of the Laws. Of the fourteen industries designated under the 1978 law, seven were authorized to form legally binding capacity reduction cartels. In each case several rounds of capacity reduction negotiations took place among firms. The dimensions of the capacity reduction programs are summarized in Table A4.1. On average, industries agreed to a 24% reduction of 1977 capacity. The planned reductions roughly equalled underutilized capacity in highly oligopolistic industries, but represented a substantially smaller share of underutilized capacity in more competitive ones. According to government reports, 12 of the 14 industries disposed of at least 90% of the capacity targeted by the stabilization plan.[3/] Importantly, in fact, under market pressure, the more competitive industries generally exceeded their capacity reduction targets.

A4.05 The economic effects of the capacity reduction were mixed. In general, industry finances improved. For example, in a sample of the designated industries, the balance of outstanding bank loans declined by 31% between 1977 and 1982. This improvement came at the expense of consumers; between 1975 and 1978 the average price of the designated industries declined by only 3%, and between 1978 and 1981 it rose by 9%.[4/] Given the apparently positive effect of stabilization programs on prices, and presumably profits, and the fact that the industries involved tend to produce homogeneous, extensively traded products, one has to assume that these output adjustments took place behind some form of protection. Generally in the case of a declining, import-competing industry, a significant price increase cannot occur in the absence of some formal or informal trade barrier. In the case of an export industry, it implies some cartelization of foreign markets. Capacity reduction was also accompanied by reductions in employment on the order of 15%. Surprisingly, about 70% of these workers were reported to have become unemployed for at least some period of time, despite tight labor market conditions, and intense efforts by Japanese firms to place workers with other establishments or other firms.

3/ Those familiar with the negotiating process suggest that most of the capacity reduction programs affected capacity proportionally across firms. In some cases the programs allocated half of the reduction target in proportion to capacity, and another half in proportion to underutilized capacity as of a date prior to the beginning of the program. In general, "side payments" were not made, and thus there was no mechanism for concentrating cutbacks on firms that owned relatively inefficient facilities. One exception is shipbuilding where an industry-wide Stabilization Association was formed independent of legislation to buy out and eventually close smaller, inefficient shipyards. Another exception is the Joint Scrapping Plan of the textile industry, where interest free government loans were used to buy out smaller companies.

4/ See Peck, et. al. (1985).

Table A4.1: CAPACITY REDUCTIONS UNDER THE 1978 JAPANESE RESTRUCTURING LAW

Industry	Tons of capacity before reduction	Goal of percent of capacity	Goal of percent of excess capacity	Net reduction as percent of goal
Concentrated Industries				
Aluminum smelting	1,642	57	210	97
Nylon filament /a	367	20	96	83
Polyester staple /a	398	20	84	76
Polyacrylonitrile staple /a	431	20	100	92
Urea*	3,985	45	90	93
Polyester filament /a	350	13	55	35
Unconcentrated Industries				
Ammonia*	4,559	26	38	100
Ferrosilicon	487	20	45	164
Shipbuilding	9,770	35	24	105
Linerboard /a	7,549	15	27	93
Phosphoric acid	934	20	42	91
Wool /a	182	12	36	236
Cotton spinning /a	1,204	6	29	136
Electric furnace	20,790	14	37	-

/a Industries with formal capacity reduction cartel.

Source: Peck et al. (1985).

A4.06 A retrospective analysis of the effects of the 19833 law is not possible because its implementation is not scheduled to be completed until 1988. A listing of the industries designated and the capacity reductions planned and already achieved appears in Table A4.2. Importantly, about half of the industries designated are "recidivists" from the 1978 law. Approximately half of the industries designated under the 1983 law have taken advantage of the provision to form business tie-ups.

A4.07 An Assessment. Essentially, a capacity reduction cartel raises prices to alleviate the financial distress faced by an industry. This transfer from the consumer to the producer may be justified on distributional grounds. Another line of argument is that the tendency of Japanese firms to engage in "excessive competition" rather than orderly private capacity reductions or the absence of adequate markets and institutions for private mergers

Table A4.2: CAPACITY REDUCTION PLANS UNDER THE 1983 JAPANESE RESTRUCTURING LAW

	Disposal goal as % of capacity	Disposal as % of goal by 1985	Capacity utilization in 1984
Electric furnace@	14	0	87
Aluminum refining@	50	0	39
Nylon filament@	(Disposal program completed)		87
Acrylic staple@	(New investment regulated	)	93
Polyester filament@	(until 06/30/86	)	78
Polyester staple@	(	)	92
Viscose staple	15	21	85
Ammonia@	20	83	74
Urea@	36	23	62
Wet phosphoric acid@*	17	62	71
Solubole phosphoric	32	46	45
Chemical fertilizer*	13	44	67
Ferrosilicon@	14	100	53
High-carbon ferrochromium	10	0	50
Ferronickel	12	0	62
Paper*	11	28	86
Corrugated board paper@	20	12	66
Ethylene*	36	74	93
Polyolefin*	22	74	100
Vinyl chloride resin*	24	90	96
Ethylene oxide*	27	61	86
Styrene	26	0	79
Rigid PVC*	18	100	76
Sugar	26	60	65
Cement	22	85	60
Electric wire	14	0	70

Legend: * = Industries participating in a business tie-up.
@ = Industries continuing from the 1978 law.

Source: Industrial Development Bank of Japan.

and acquisitions.[5/] From a distributional viewpoint, the price increase engineered by a capacity reduction cartel typically reverses a steep decline in prices due to world market conditions. These cartels, therefore, tend to prevent, rather than create, windfall transfers between consumers and producers. While the cartel-induced transfer could in theory be targeted to ease the specific income dislocations caused by the change in world prices, the evidence suggests that Japanese transfers primarily compensated creditors and not employees. From an efficiency perspective, it is very likely that intervention raised prices above marginal production cost and generated inefficiency. While the scope of this inefficiency may be modest in the short run, long-run losses are more significant, especially since cartelized price levels can be maintained for several successive five year periods. Routine rescue efforts tend to discourage capacity reduction, particulatly by smaller firms, since the cartels generally allocate production targets irrespective of efficiency. Also, the program creates "moral hazard" among lenders, who may bear much larger risks than they would in the absence of a government rescue program.[6/]

A4.08 Implications for Korea Policy. Japan's industrial adjustment policies have enjoyed wide domestic political support. In sharp contrast to the adjustment efforts of many other OECD countries. To be sure, Japan's adjustment problems were more modest, and occurred in the context of rapid growth. Still, the Japanese experience is of great interest to Korea, both because of its supposed success and because of other parallels between Japanese and Korean development patterns. The Japanese approach may have some limited reference in the medium term, that is, until Korean capital markets develop enough to provide wider options for dealing with restructuring. At the same time, there are major structural differences between the Japanese and Korean economies that preclude direct application of Japanese methods and which may impose intolerably high costs on the economy if this approach is followed.

A4.09 Most obviously, Korean domestic markets are much smaller than Japanese markets because of differences in both population size ad per capita income. Thus Korean firms are less likely to engage in unprofitable production simply to maintain long-run market share.[7/] Second, the Korean economy is more open, and Korean industries depend on foreign markets to a far greater extent than Japanese industries. Third, in Korea, transfers from

---

5/ Indeed, mergers are very rare in Japan, but this observation alone does not prove that capital markets are flawed. In any case, capital market imperfections are more likely to call for capital market interventions than government intervention in industrial structure.

6/ Ultimately, these rescues prove too costly, in particular if market signals are ignored. Thus, for example, Japan recently permitted the bankruptcy of the nation's largest shipping firm.

7/ Relative to Japan, Korean industry is less capital-intensive, and there is no equivalent of Japanese lifetime employment.

domestic consumers could at best represent a small fraction of the financing requirements of industries in distress. Furthermore, the trade barrier required to implement domestic cartels are likely to elicit external pressure from Korea's trade partners.

A4.10 Perhaps because of differences in economic size, the financial "mix" of Japanese restructuring programs emphasizes transfers from consumers (through the formation of cartels and deemphasizes direct financial support from government. The absence of government financing is desirable, but it hinges on the close relationship between Japanese firms and their banks, which are often members of the same group, and which in some cases own a substantial share of the troubled company. To the extent that the Korean private sector becomes involved in restructuring, it would directly involve financial support which exacerbates the problem of moral hazard, retards the process of financial liberalization, and perpetuates Government's role in industrial decision-making. Moreover, the inefficiencies inherent in increasing the prices of intermediate goods can be especially severe in Korea, as they impair the profitability of up-stream export industries. For all of these reasons, the Japanes adjustment model, despite is apparent attractions, has limited applicability in the Korean context.

CHAPTER 5: THE ROLE OF FINANCE AND INDUSTRIAL PERFORMANCE

A. Introduction

5.01 Korea started an ambitious economic development and industrialization plan in the early 1960s. In the process, the Korean Government made extensive and forceful use of a wide range of incentives designed to assure private industry's close compliances with its plans. Among them, probably the most widely used has been differential access to credit from commercial banks as well as specialized banks under government control. In addition, foreign loans which were an attractive source of borrowing required government authorization and its allocation was largely controlled by the government. Availability of credit was, and continues to be, an extremely effective instrument because of the highly leveraged nature of Korean industry and the implicit subsidy inherent in preferential credit which had to be rationed. This was accomplished essentially by government directive.

5.02 Under these circumstances, access to a stable and low-cost source of credit, e.g., bank loans, was considered to be crucial for firm performance. Being disfavored by Government, which owned the banks and controlled their loans, could easily lead to a firm's demise, particularly since firms are heavily indebted, and, furthermore, because about two thirds of their liability is in the form of short-term debt. In essence, the Korean Government provided two incentives for the development of strategic industries which had an important bearing on private firms' investment decisions. First, the Government changed the expected rate of return of investment by controlling the interest rate and allocation of bank loans. By favorably allocating cheap bank credit to a strategic sector, the Government could reduce the cost of investment and increase the expected rate of return in that sector. Second, the Government could change the perceived risk of investment by assuring a stable flow of bank loans to selected industries regardless of their economic or financial performance. Thus, if a firm followed government policy by investing in projects which had government support, its bankruptcy risk was reduced, perhaps even eliminated. In other words, the Government, by controlling the financial sector, became a risk partner of firms, encouraging them to undertake projects which might have been declined otherwise. Therefore, in the case of Korea, it would not be very meaningful to discuss industrial development without looking at the role of government in financial allocation.

5.03 It is not easy, however, to evaluate the role of finance in the industrial development and performance. The performance of a firm or an industry is affected not only by factor market conditions, such as financial cost and labor cost, but also by product market conditions such as the degree of competition, market structure and external and internal demand conditions. Furthermore, it is affected by other government policies with respect to trade, foreign exchange, pricing and taxes, among others. In these circumstances, it is difficult to separate out the effect of particular factors. Most studies on industrial performance and structure assume that factor markets are competitive and neutral.[1/] Clearly, however, this has not been the case in Korea, where Government's intervention in factor markets (viz., the financial market) was so extensive that it may as well have affected the performance of industry as much as product market conditions did.

5.04 Therefore, an attempt is made here to investigate the role of government-directed finance on the development of industry since, in the Korean context, access to concessionary credit has been a primary signal of strategic industry policy on the part of Government and also has been an important factor for firm performance. In the 1970s, foreign and bank loans, the allocation of which was controlled by Government, accounted for about half the source of funds of the corporate sector and the costs of these two sources of credit were the cheapest among various sources of borrowing due to the strict interest rate ceilings on the bank credit and continuously fixed exchange rates.[2/] Therefore, firms which had better access to those sources of borrowing, especially to bank policy loans, which were much lower cost than the already highly subsidized general bank loan, faced a lower average cost of capital.[3/]

5.05 The approach taken in this chapter is to assess the degree to which differential access to credit by the different sectors of the economy was created by Government's control of credit allocation and the difference in the average cost of capital paid during the sample period, 1972-84. In conjunction with this, the question of whether the allocation of government-controlled credit was efficient or not will be examined, to a limited extent, by comparing the access to credit of different sectors with the average rate of return on the investment in the sectors. The approach also looks at whether Government favoritism in terms of credit allocation has affected the performance of particular sectors and industries. Furthermore, it addresses the issue of how financial liberalization in the 1980s improved the allocation of credit between different sectors or industries compared to that of the 1970s.

5.06 The measures examined include the ratios of (i) total bank loans and foreign loans over total assets was chosen as a measure of access to Government-controlled credit, (ii) financial expense (interest rate paid plus discount) over total borrowing as a measure of average cost of borrowing, (iii) profit before subtracting financial expense over total asset (capital stock) as a measure of average rate of return on capital investment, and (iv) normal ratio, as well as (v) net sales growth rate adjusted for inflation

---

1/ In other words, in the literature of industrial economics, firms usually compete equally in the factor markets and factor markets just take the role of passively meeting the firm's demand for input which again is dependent on the output market conditions.

2/ See Tables 5.3 and 5.4 for the measure of implicit subsidy that accompanied the bank and foreign loan allocation in the 1970s.

3/ See Table 5.5 for the relative size of policy loans.

as measures of performance of each industry. The data originate from the Bank of Korea's (BOK) Financial Statement Analysis, which reports aggregated balance sheets and income statements by industry and sector, and includes 92% of total corporate sales, although only 45% of total firms.

5.07 The sectors compared here are (i) large firms vs. small firms, (ii) export industry vs. domestic industry, and (iii) heavy and chemical industries (HCI) vs. light industries, following the BOK classification.[4/] In Korea, as was mentioned, the sectors which received the highest government priority in the 1970s were export and HC industries. In the process of severe credit rationing under low interest rate ceilings, large firms were favored over small and medium firms because of the low transaction costs and perceived risk to bankers.

B. The Sectoral Experience

5.08 The analysis of the period of 1972-84 yielded the following conclusions. In the 1970s, there were substantial gaps in the access to capital with favored sectors having greater access to subsidized sources of capital (bank loans and foreign loans) than others (Table 5.7). This seems to have contributed to their rapid growth and high profitability despite their relatively low rates of return. The Government's strong intervention seems to have created substantial distortions in the allocation of financial resources by creating large disparities in the average cost of capital among different sectors. It is to be expected that substantial gaps in the marginal rates of return on investment across sectors should have resulted. Large firms, the export industry and HCI had greater access to capital and, as a result, had lower capital costs than did small/medium firms, domestic industry and light industry, respectively, in the 1970s (Table 5.8). Although the former's rates of return on investment were significantly lower, indicative perhaps of capital underutilization, they did exhibit higher growth rates (Tables 5.10-5.12).

5.09 In the 1980s, however, the gaps in accessibility to and cost of capital have been greatly reduced, or in some cases (small/medium vs. large firms, for example) completely eliminated as a consequence of the financial

---

4/ According to the Financial Statements Analysis, large firms are those which employ more than 300 employees in the case of manufacturing firms, more than 200 in the case of the service industry. HCI includes: industrial chemicals (351), petroleum refineries (353), other nonmetallic mineral products (369), basic metals (37), fabricated metal products (381), machinery (382), electrical, electronic machinery and appliances (383), transportation equipment (384), and precision equipment (385). A firm is classified into export industry if export constitutes 50% or more of its total sales. Otherwise, it is classified as domestic industry. Therefore, the classification of domestic industry vs. export industry is not as clear as other classifications. The figures in parentheses are the code numbers of the Korean Standard Industry Classification.

liberalization effort in this period. The shrinking share of government-controlled credit (bank loans and foreign loans) in corporate finance and the explosive expansion of credit from nonbanking sectors (including the securities market), in which the cost of borrowing is much freer of government intervention, contributed to more equalized cost of and access to capital in the 1980s. At the same time, some progress in the liberalization of the banking sector, more importantly the elimination of the interest rate gap between general loans and various policy loans, has contributed to reducing the disparity in the cost of capital, and, as a result, contributed to reduce distortions in the financial allocation. Changes in government credit allocation policy to correct past misallocations and favoritism also contributed to better allocation. Comparisons of access to capital by different sectors and their relative performance are described below, with particular attention paid to the protection cum repression period of the 1970s versus the liberalized period of the 1980s.

Large Firms vs. Small/Medium Firms

5.10 Large firms had relatively much greater access to preferred credit than did small/medium firms during most of the period (Table 5.7). The gap was sizable during the 1972-76 period, began to narrow over the 1977-81 period and was reversed thereafter. Since 1982, small/medium firms have enjoyed better access to borrowing from banks and abroad than have large firms. This reversal seems due to (i) a reduced share of foreign borrowing on the part of large firms, (ii) government policy to increase the bank lending to small/medium firms, and (iii) increased real interest rate which gave rise to less bank rationing in favor of large firms. As a result of greater access to subsidized credit, large firms faced a lower average cost of capital than did small/medium firms in the 1970s (Table 5.9). The disparity in borrowing cost was consistently present over the 1972-78 period, and was sizable on a percentage basis. Abruptly in 1979, this disparity was eliminated and has hovered at close to neutrality since then. Greater access to and lower cost of capital seems to have enabled the large firms to grow faster than small/medium firms, despite the former's relatively low rate of return on investment (Tables 5.10-5.12).

Heavy Chemical Industry vs. Light Industry

5.11 A similar pattern result is observed in the comparison between HCI vs. light industry: HCI had greater access to capital, especially in the latter half of 1970s during which Government continued to promote HCI development (Table 5.7).[5/] As a result of substantial access to preferential

---

5/ HCI's advantage in access to capital seems to be greater than the data in Table 5.7 suggests if we consider that the HCI is leveraged to a lesser extent than light industry (Table 5.8) due to the higher proportion of its fixed assets which are financed through equity. This is, the proportion of bank and foreign loans in total borrowing (net assets) should be much higher for HCI compared to light industry than the figures in Table 5.7 suggest.

borrowing,[6/] the cost of capital to the HCI sector was considerably below that for light industry in the same period (Table 5.9), by as much as 20% to 35% as seen in Figure 2.3. This great advantage in terms of access to and cost of capital, among other things, seems to have contributed to HCI's rapid growth in the 1970s. It grew faster than light industry despite the fact that it had much lower investment efficiency compared to the latter (Tables 5.10 and 5.12) in the 1970s. This lack of efficiency is continued by the data on capacity utilization as well as sectoral ICORs (see Tables 2.6 of Volume I and 1.11 of Volume II).

5.12 The gap in the access to and cost of capital between HC and light industries started to decrease in the 1980s. The elimination of preferential lending rates for policy loans in 1982 and expansion of the nonbanking sector with a relative diminution in the role of bank and foreign loans seems to have contributed to reducing the gap. On the other hand, the disparity in the rates of return on investment of the HCI and light industry has been falling in the 1980s as the excess capacity of HCI has been gradually reduced as the export share of the industry increased (see Chapter 4 of Volume I). As of 1984, there was no significant difference in the cost of capital and return on investment measures between the two, which provides some evidence that the efficiency of capital allocation between the two sectors has much improved compared to the 1970s.

Export Industry vs. Domestic Industry

5.13 There was also a substantial gap in the access to credit between export and domestic industries over the period, with the former exhibiting greater access, and, as a result, experiencing lower costs than the domestic industry during most of the period. Interestingly, its advantage vis-a-vis domestic industry in terms of access to and cost of capital was biggest during 1980-82. This may be partly explained by the impact of automatic access to export-related loans under conditions of tight credit control during this period. The rate of return on investment for the export industry was much lower than that for the domestic industry during 1974-79. This seems to be greatly influenced by the fixed exchange rate policy of the period which led to overvaluation (see Table 5.4), which appears to have offset the favorable impact of cheaper credit.

5.14 The growth of exports was, on average, lower than the growth of domestic industry in the 1970s, although there were ups and downs, reflecting the internal and external environment. In a sense, export credit in the 1970s was partially neutralizing the negative effect of high protection and overvaluation on the export industry. Without favored access to borrowing at subsidized rates, the performance of the export industry could have been much poorer indeed. Or putting it differently, the export industry faced

---

6/ According to the KDI, 93% of policy loans (excluding export loans and small/medium firms' promotion funds) was given to HCI and only 7% was given to light industry in 1978. See The Issues of Industrial Policy and Its Direction (Korean), KDI, December 1982, p. 46.

disadvantages in the 1970s despite favored treatment in the credit allocation process, although, as can be seen from Figure 2.3 of Volume I, that preference was quite variable, and indeed was declining over the 1975-79 period. It took a major reversal in 1980 to restore any semblance of an export credit preference.[7/] Continuous depreciation and import liberalization in the 1980s seem to have contributed to the improved return on investment in the export industry and its profitability in the 1980s (see Tables 5.10 and 5.11). As of 1984, the rates of return on investment were almost equalized between the two sectors.

C. The Experiences of Specific Industries

5.15 Similar results are observed when the specific industries cases are studied. In general, manufacturing, electricity, fishing and transportation industries had good access to subsidized credit and consequently paid relatively less for capital. On the other hand, wholesale and retail trade, mining, and real estate and business service industries had poor access to borrowing and faced relatively high costs of capital (see Table 5.13). The construction industry has increased its access to bank and foreign loans in the 1980s, due to exceptional circumstances, while all other industries' share of foreign and bank loans over total assets have been falling. Among manufacturing industries, the HC industries and the export industries benefitted from good access to borrowing while light industries, especially the industries which produce mainly for the domestic market, exhibited poor access. Textile and apparels, and wood and furniture industries, which were the major export industries in the 1970s, had favorable access to capital while the food and beverages, and paper and printing industries, which basically produce for domestic market, had relatively poor access (Table 5.13). Consequently, the latter had significantly higher costs of capital than did the former (Table 5.15). Among HC industries, basic metal industry, which includes iron and steel industry, had rapidly increasing access to borrowing from banks and abroad in the latter half of the 1970s, suggesting that it attracted a large part of new loans during this period. Fabricated metal product and equipment industry (which includes shipbuilding and automobile industries) also had good access to borrowing in the latter half of the 1970s. Consequently, these industries, especially basic metal industry, paid much less for capital than did on average manufacturing industries.

5.16 Although it is not possible to establish an exact linkage between access to capital, the resulting cost, and the relative performance of different industries, the analysis of data strongly suggests that the government-controlled access to and cost of credit affected the profitability and growth of specific industries. The fishing industry, for example, was able to secure access to bank loans in the 1970s, while its profit rate was negative for most of the period (Table 5.17). Without cheap bank credit, therefore, this industry may have declined faster. The electricity industry, despite its very low rate of return on investment relative to other

---

7/ See Rhee Sung Sup (1985) for similar findings on the decline of incentives to exports.

industries, enjoyed high profitability, largely due to its access to bank and foreign loans and resulting cheap cost of capital. Service industries (wholesale and retail, and hotel, real estate and business service, and construction) had very high rates of return on investment relative to other industries in the 1970s (Table 5.16). However, their access to capital was quite poor in the 1970s. If the credit market had not been strongly controlled, the growth of these industries may have been faster and its current share in Korean industry might have been higher. Similar arguments can be made for industries in the manufacturing sector. Without intervention in credit allocation, some of the HC industries such as the basic metal industry would not have obtained such cheap access to funds (see Table 5.16 for its relatively low rate of return in the 1970s) and may not have grown as fast. On the contrary, it may be said that paper and printing, and food and beverage industries may have grown faster had financial allocation been determined in the free and competitive market.

D. Overall Assessment

5.17 The Korean Government's intervention in the financial sector seems to have been quite distortive, especially in the latter half of the 1970s when it strengthened the control over credit allocation to favor the development of HC industries, although the interventions in the 1960s and 1970s to support export activities may have had positive aspects in the sense that they neutralized the distortive effects of other government policies such as high protection and overvaluation. By creating wedges between the average cost of capital among various sectors, public policy helped produce disparities in the marginal rates of return to capital. In addition, the Government's strong control over the financial sector affected firm behavior and may have promoted decision-making patterns, now deeply rooted, which were not conducive to the efficient use of resources in the economy. In general, the strong government intervention in the financial sector during the last two decades seems to have affected the following developments which have deeply permeated through the patterns of industrial development in Korea.

Distortive Allocation of Credit

5.18 The government-controlled financial allocation gave rise to a high distortion in the allocation of capital among various sectors by increasing the gap in the cost of capital, and consequently the gaps in marginal rate of return of capital between the favored sector and disfavored sector. As a result, it is suspected that the resulting allocation of capital could not maximize the growth potential of the country.[8/] The data suggest that a higher proportion of bank credit went to less productive borrowers, or the borrowers who could get much of bank loan with subsidized cost expanded their business or production lines in a way too much capital-intensive and consequently low marginal rate of return on capital given the endowment of

---

8/ A qualification of this argument is that product market distortions were not as significant as the distortions in the financial cost among various sectors.

capital and labor of the economy (see large firms vs. small firms, for example). The significant gaps in the cost of capital and the rate of return on the capital among different sectors and industry suggest that the overall efficiency of capital allocation could have been increased if the financial allocations had been determined by the more market-oriented financial system.

High Leverage of Firms and Poor Development of the Equity Market

5.19 The cheap cost of debt due to interest rate ceilings made debt finance quite attractive to the corporate sector. The real cost of bank loans, a major source of corporate debt, was negative throughout most of the 1970s. In addition, Government's de facto risk partnership with the corporate sector, exercised through its control over the banking institution, increased corporate access to debt finance even further. Past government actions to avoid explicit bankruptcies of large firms in favored industries created the perception that debt finance was not very risky compared to equity finance. As a result, the corporate debt ratio became increasingly high in the 1970s, and development of the equity market was slow despite government efforts to the contrary (see Table 5.18). Paradoxically those who could have raised capital through the equity market, the large and well-established firms, had no incentive to do so because of their preferred borrowing position. The ensuring high debt ratios made the corporate sector as a whole very vulnerable to external shocks and economic fluctuation, thus prompting more government involvement in the banking system, and its credit allocation, to bail out troubled firms and industries.

High Industrial Concentration

5.20 The financial repression and the resulting credit rationing tended to favor large firms because of low transaction costs involved. The best thing the banks could do to increase their profits in the presence of strict interest rate ceilings and undifferentiated (for risk) interest rates was to reduce the transaction cost and default risk. This led them to favor large firms in those sectors favored by government. This resulted in higher loan concentration and eventually higher industrial concentration (See Chapter 1).

Slow Development of the Financial Sector

5.21 Heavy government financial intervention interfered with the development of the financial sector not only with respect to its growth, but also with respect to quality of its services. Intervention in the banking industry (i.e., in its asset management and day-to-day operations) removed any incentive for the banks to innovate in their operations or to become more efficient in the intermediation of financial resources. Nor did the banks have much incentive to select profitable borrowers because they were not rewarded for doing so. Rather, banks passively accommodated the credit demands of the government-favored borrowers. The repressed interest rates which fluctuated around zero or were negative in real terms made bank deposits quite unattractive financial assets and made the holders of bank deposits the ultimate bearer of the cost of financial intervention. Consequently, the financial sector's growth was more sluggish than it would otherwise have been. Furthermore, in the process of channeling government-directed loans, the commercial banks accumulated a substantial amount of nonperforming loans which reduced profitability and limited its future development.

The Implications of Financial Liberalization Since 1980

5.22 The financial liberalization efforts since 1980 have greatly improved various aspects of financial allocation. First, it reduced the gaps in the access to and cost of capital between different sectors as we see in Tables 5.7 and 5.9. By reducing these gaps, the financial allocation since 1980 became less distortive. The small/medium firms could get better access to borrowing and this has resulted in more equalized cost of borrowing between large firms and small firms. Owing to this improved access to capital, small/medium firms since 1980 have been growing faster than large firms which seems to be desirable in the sense that the latter's rate of return to investment is higher. This may well indicate that overall allocative efficiency has increased. It will also help to reduce the market concentration and encourage more competition in the domestic market.

5.23 The gap in the cost of borrowing between the export industry and the domestic industry also has been reduced since 1982 as the preferential lending rates were abolished, although the gaps in access still exist. Similarly, the gaps in the cost of and access to capital have been significantly reduced between HC industry and light industry. All these suggest that the financial allocation since 1980, especially since 1982, became much less distortive, compared to that of the 1970s. The high real cost of debt along with the reduced degree of (implicit) government commitment to be a risk partner in the case of bad performance of industrial firms also pushed firms to reduce their dependence on debt. As a result, the debt ratio has been decreasing rapidly since 1980 which seems to be desirable not only for the stability of the corporate sector and lower vulnerability to with respect to external shock, but also for further liberalization of the financial system.[9/]

E. Future Implications

The Banking Industry

5.24 The government's current financial policy is characterized by an emphasis on small/medium firm financing, a reduced commitment to financial support for HC industry, and the abolishment of various preferential lending

---

9/ Financial liberalization, grounded on decelerating inflation, also contributed to a rapid expansion of the financial sector, especially less regulated nonbanking financial institutions. With more savings channeled through financial institutions which became less regulated, it is expected that the overall capital allocation became more efficient and the intermediation cost of capital in the economy has decreased.

rates on policy loan (including export loans).[10/] It seems to have contributed to correcting distortions in the financial allocation caused by previously heavy financial repression. One area in which Government has not pursued liberalization as strongly as its other efforts seems to be vis-a-vis its commitment to bailing out troubled firms. Since the relief loans of banks and distress borrowing of industrial firms put a significant yoke on the development of the banking industry and further liberalization of the financial sector, it may be desirable to reduce the burden on the bank's asset management by reducing support for some inefficient firms and helping to extricate the banks. Although this may give rise to temporary shocks in the economy, this type of policy will also facilitate the industrial restructuring of the economy and reduce the structural adjustment costs in the long run.

Capital Market

5.25 If the process of further financial liberalization is to involve a decreasing role of Government as a risk partner of the industrial sector, it would be advisable to seek a substitute source of risk capital. Of course, this is particularly tricky in the highly leveraged Korean industrial environment. In the case of the United Kingdom and the United States, a well-functioning equity market and merchant banks provide risk capital to the corporate sector. In Germany and France, universal banks and industrial banks provide risk capital and become risk partners of the industrial sector. In Japan, the group banks provide loans to the group firms and become risk partners. As the Korean economy becomes more industrialized, and its banking system more liberalized, the capital market which could provide the risk capital to industrial firms needs further development.

Stable Inflation

5.26 It also seems desirable for the stable growth of the industrial as well as the financial sector to maintain the current low rate of inflation. Korean firms' profitability is, as we see in Table A5.19, quite sensitive not only to the level of nominal but also real interest rates. Of course, inappropriate inflation accounting may have underestimated the profitability of firms which are net debtors in the inflationary period. However, it is true that a substantial part of corporate debt will be held in the form of liquid financial assets which cannot be fully hedged against inflation (such as demand deposit or cash) and firms also suffer from inflation as their costs (financial expenses) increase faster than its revenue if the nominal interest rate increases as fast as inflation. High real rates are necessary to encourage financial saving whereas high real rates are a serious problem for the industrial sector which must roll over a high level of debt. As long as inflationary expectations can be contained, interest rates can be brought down somewhat without a loss in savings generation but with a considerable gain for firm profitability and financial stability.

10/ This does not imply, however, that credit rationing does not continue to be necessary with interest rates in the banking sector still under government control.

Interest Rates and Corporate Profitability

5.27 If the current low level of inflation continues, and if peoples' expectations settle down, the Government may be able to reduce slightly the current nominal interest rates at some point in the future to boost the profitability of industrial firms. Currently, the corporate sector seems to be confronted with historically high real interest rates, although it is not possible to judge what would be the optimal level of interest rate in the absence of controls. If we compare the cost of borrowings and profitability of Japanese firms in the 1970s (see Table A5.20 of Volume I), Korean firms are now paying higher real interest rates and their profitability is lower. However, we have to note that, if the Government reduces interest rates prematurely, it may risk retarding financial sector growth.

Policy Coordination

5.28 Finally, it cannot be overemphasized that the financial liberalization policy that the Government is pursuing now should be well coordinated with other government policies. Even if financial distortions are small due to relaxed regulations on the financial systems, the system may be allocating resources to economically inefficient projects if the price and profitability signals it receives are incorrect. Therefore, in order to maximize the growth potential of an economy, it is very important to have a consistent liberalization strategy which fully integrates macroeconomic policy, trade policy, pricing policy and financial sector policy.

Table 5.1: SOURCES OF NIF
(Million won)

Sources	1974	1975	1976	1977	1978	1979	1980	1981	1982	1983	1984	Total
1. NIF bonds issued and deposits	71,722 (98.7)	131,688 (97.8)	165,213 (90.0)	276,129 (89.3)	307,471 (68.2)	291,901 (64.1)	372,069 (71.7)	492,152 (64.1)	353,933 (46.0)	443,760 (53.9)	73,326 (12.6)	2,979,362 (58.8)
National savings association	8,776 (12.1)	11,307 (8.4)	12,121 (6.6)	17,027 (5.5)	26,448 (5.9)	38,990 (8.6)	41,814 (8.1)	47,964 (6.2)	39,671 (5.2)	13,086 (1.6)	18,055 (3.1)	275,259 (5.4)
Public funds	11,512 (15.8)	16,771 (12.5)	25,341 (13.8)	45,368 (14.7)	54,586 (12.1)	64,381 (14.1)	91,459 (17.6)	75,597 (9.8)	37,416 (4.9)	40,046 (4.9)	189,498 (32.5)	277,979 (5.4)
National life insurance and postal savings	10,046 (13.8)	15,000 (11.1)	18,500 (10.1)	-2,089 (0.7)	- -	- -	- -	-24,275 (3.2)	-17,184 (2.2)	- -	- -	- -
Banking institutions	38,746 (53.3)	81,153 (60.2)	97,044 (52.8)	197,769 (64.0)	202,359 (44.9)	172,281 (37.8)	214,768 (41.4)	372,200 (48.5)	286,064 (37.1)	275,129 (33.4)	158,102 (27.1)	2,095,614 (41.3)
Insurance companies	2,642 (3.6)	7,457 (5.5)	12,207 (6.6)	18,054 (5.8)	24,078 (5.3)	16,249 (3.6)	24,029 (4.6)	20,664 (2.7)	7,966 (1.0)	115,449 (14.0)	86,667 (14.9)	335,511 (6.6)
Nonlife insurance companies	2,642 (3.6)	7,457 (5.5)	12,207 (6.6)	18,054 (5.8)	24,078 (5.3)	16,249 (3.6)	24,028 (4.6)	20,664 (2.7)	7,966 (1.0)	42,290 (5.1)	1,002 (0.2)	174,633 (3.4)
Life insurance companies	- -	- -	- -	- -	- -	- -	- -	- -	- -	73,209 (8.9)	87,669 (15.1)	160,878 (3.2)
2. Collections of loans	954 (1.3)	5,712 (4.2)	7,469 (4.1)	27,987 (9.7)	75,166 (16.7)	92,801 (20.4)	131,365 (25.3)	195,242 (25.4)	254,675 (33.1)	338,793 (41.2)	419,280 (72.0)	1,549,444 (30.6)
3. Carryover from previous year	- -	2,731 (2.0)	10,963 (6.0)	5,103 (1.7)	68,482 (15.2)	70,820 (15.5)	15,712 (3.0)	80,730 (10.5)	164,428 (21.0)	40,033 (4.9)	89,995 (15.4)	540,535 (10.7)
Total	72,676	134,669	183,645	309,219	451,119	519,146	768,124	770,036	822,586	582,601	506,934	

Note: ( ) = share.

Table 5.2: USES OF NIF
(Million won)

Uses	1974	1975	1976	1977	1978	1979	1980	1981	1982	1983	1984	Total
Heavy and chemical industries	44,120 (58.5)	61,348 (49.6)	100,677 (56.4)	147,218 (61.2)	239,447 (62.7)	285,299 (64.9)	266,704 (60.8)	332,997 (54.9)	445,003 (61.0)	472,636 (64.5)	318,708 (65.2)	2,714,158 (61.1)
Agriculture	10,547 (14.0)	11,400 (9.2)	19,415 (10.9)	9,998 (4.2)	14,946 (13.9)	19,835 (4.5)	21,711 (5.0)	38,698 (6.4)	45,000 (6.2)	44,954 (6.1)	30,000 (6.1)	266,506 (6.0)
Power industries	17,000 (22.5)	43,200 (34.9)	40,000 (22.4)	50,000 (20.8)	92,000 (24.2)	100,000 (22.7)	120,000 (27.4)	160,000 (26.4)	140,000 (19.2)	40,000 (5.5)	50,000 (10.2)	852,000 (19.2)
Saemaul factories	3,740 (5.0)	4,757 (3.8)	8,450 (4.7)	3,521 (1.5)	4,222 (1.5)	4,676 (1.1)	- -	- -	- -	- -	- -	29,365 (6.6)
Export on deferred payment	- -	3,000 (2.4)	10,000 (5.6)	30,000 (12.5)	29,684 (6.2)	30,000 (6.8)	30,000 (6.8)	75,000 (12.4)	10,000 (1.4)	175,000 (23.9)	100,000 (20.5)	582,674 (13.1)
Total (A)	75,407 (AIB)	123,706 (2.5)	178,542 (3.1)	240,738 (4.1)	380,299 (4.0)	438,810 (4.4)	438,415 (3.7)	606,696 (2.6)	730,003 (2.8)	732,590 (2.7)	498,708 (2.3)	4,444,913 (1.4)
Total Domestic Credit (B) /a		3,006	3,973	4,387	5,979	8,722	11,826	16,778	22,016	27,529	31,847	36,056

/a Billion won.

Note: Numbers in parenthesis are percents.

Table 5.3: INTEREST RATES ON VARIOUS LOANS
(%)

Year	Curb market /a	Corporate bond	Bank loan				Inflation (GNP Deflator)
			General	Selected policy loan			
				Export	MIPF /b	NIF /c	
1971	46.41	-	22.0	6.0	-	-	13.92
1972	38.97	-	19.0	6.0	-	-	16.11
1973	33.30	-	15.5	7.0	10.0	-	13.40
1974	40.56	-	15.5	9.0	12.0	12.0	29.54
1975	41.31	20.1	15.5	9.0	12.0	12.0	25.73
1976	40.47	20.4	17.0	8.0	13.0	14.0	20.73
1977	38.07	20.1	15.0	8.0	13.0	14.0	15.67
1978	41.22	21.1	18.5	9.0	15.0	16.0	21.39
1979	42.39	26.7	18.5	9.0	15.0	16.0	21.20
1980	44.94	30.1	24.5	15.0	20.0	22.0	25.60
1981	35.25	24.4	17.5-18.0	15.0	11.0	16.5-17.5	15.90
1982 Jan	32.64	17.29	15.5-16.0	12.0	15.0	15.5-16.5	13.20
Mar	32.60	17.29	13.5-14.0	11.0	13.5-14.5	13.5-14.5	13.20
Jun	33.12	17.29	10.0	10.0	10.0	10.0	7.60
1983	25.77	14.23	10.0	10.0	10.0	10.0	3.00
1984	24.84	14.12	10.0-11.5	10.0	10.0-11.5	10.0-11.5	3.90
1985	24.0	14.2	10.0-13.0	10.0	10.0-11.5	10.0-11.5	3.50

/a Source: BOK.
/b Machinery Industry Promotion Fund.
/c National Investment Fund.

Table 5.4: EFFECTIVE COST OF FOREIGN LOANS

Year (Dec-Dec)	Exchange Rate	Rate of change of E.R.	Interest cost of domestic Bank loan (A)	Foreign Interest Cost Nominal/a	Foreign Interest Cost Effective/b (B)	Gap (A) - (B)
1968 Dec	281.50	8.00	25.2	7.13	15.13	10.07
1969 Dec	304.45	4.00	24.0	10.06	14.06	10.06
1970 Dec	316.65	17.85	24.0	6.75	24.60	-6.00
1971 Dec	373.20	6.88	22.0	5.81	12.69	9.31
1972 Dec	398.90	-1.00	15.5	6.19	5.19	10.31
1973 Dec	397.50	21.76	15.5	10.03	31.79	-16.29
1974 Dec	484.00	0	15.5	10.19	10.19	5.31
1975 Dec	484.00	0	15.5	6.63	6.63	8.87
1976 Dec	484.00	0	17.0	5.38	5.38	11.62
1977 Dec	484.00	0	15.0	7.50	7.50	7.50
1978 Dec	484.00	0	18.5	12.31	12.31	6.19
1979 Dec	484.00	36.34	18.5	14.44	50.78	-32.28
1980 Dec	659.90	6.1	19.5	16.75	22.85	-3.35
1981 Dec	700.50	6.78	16.5	14.81	21.59	-5.09
1982 Dec	748.00	6.35	10.0	9.50	15.85	-5.85
1983 Dec	795.50	3.79	10.0	10.06	13.85	-3.85
1984 Dec	825.70	7.6	11.0	10.75	18.35	-7.35
1985 Dec	888.45					

/a Libor-based cost of borrowing.
/b Nominal cost adjusted for annual rate of change of exchnage rate.

Source: Bank estimates.

Table 5.5: SIZE AND SHARE OF POLICY LOAN BY DMB

Year	Total	General loans		Policy loans		(Of which export loans)	
1970	722.4	510.5	(70.7)	211.9	(29.3)	55.9	(7.7)
1971	919.5	647.2	(70.4)	272.3	(29.6)	80.1	(8.7)
1972	1,198.0	846.4	(70.4)	351.6	(29.3)	108.4	(9.1)
1973	1,587.5	1,052.4	(66.3)	535.1	(33.7)	224.1	(14.1)
1974	2,427.8	1,637.2	(67.4)	790.6	(32.6)	360.2	(14.8)
1975	2,905.5	2,117.9	(72.9)	787.6	(27.1)	339.2	(11.7)
1976	3,724.9	2,497.9	(67.1)	1,227.0	(32.9)	461.8	(12.4)
1977	4,709.0	3,115.4	(66.2)	1,593.6	(33.8)	567.4	(12.0)
1978	6,609.0	4,154.7	(62.9)	2,454.3	(37.1)	883.2	(13.4)
1979	8,977.8	5,827.1	(64.9)	3,150.7	(35.1)	1,227.2	(13.7)
1980	12,204.4	7,904.8	(64.8)	4,299.6	(35.2)	1,720.8	(14.1)
1981	15,955.0	10,585.6	(66.3)	5,369.4	(33.7)	2,197.2	(13.8)
1982	20,225.8	13,801.6	(68.2)	6,424.2	(31.8)	2,278.4	(11.3)
1983	24,150.3	16,190.8	(67.0)	7,959.5	(33.0)	2,620.0	(10.8)
1984	27,978.9	19,101.6	(68.3)	8,877.3	(31.7)	2,765.4	(9.9)
1985	33,810.7	23,382.6	(69.2)	10,428.1	(30.8)	3,129.9	(9.3)

Source: Bank of Korea.

Table 5.6: THE USES OF COMMERCIAL BANKS NET INCREASE OF DEPOSIT (1981)
(100 million won)

	Amount	Share
Net deposit increase (A)	16,664	100.0
Directed uses (B)	9,445	56.7
Reserve requirement	917	5.5
Contribution to NIF	2,176	13.1
Policy loan /a	6,352	38.1
(Export credit)	(780)	(4.7)
(Other)	(5,572)	(33.4)
Undirected uses (A) - (B)	7,219	43.3

Source: Korean Economy and Banking, B. J. Kim and Y. C. Park, 1984.

Table 5.7: ACCESS TO BORROWINGS BY EACH SECTOR /a
(%)

	1972	1973	1974	1975	1976	1977	1978	1979	1980	1981	1982	1983	1984
Tot Mnf	45.41	43.21	45.22	40.27	40.97	41.32	39.29	36.94	38.55	38.05	32.53	30.81	28.17
L Mnf	45.72	43.55	45.65	40.93	41.36	41.38	39.69	37.32	39.25	38.81	32.26	30.76	27.84
S Mnf	27.27	26.54	24.44	27.38	34.98	40.79	37.02	34.60	33.79	34.31	33.87	31.19	30.40
S Mnf - L Mnf	-18.45	-17.00	-21.20	-13.56	-6.38	-0.59	-2.67	-2.72	-5.46	-4.50	1.61	0.43	2.56
X Inds	47.13	45.95	49.78	45.07	43.11	44.06	42.85	41.10	48.57	45.63	38.07	35.53	32.28
Doms Inds	44.63	41.75	42.93	36.62	39.91	39.83	37.54	35.24	31.66	32.84	29.00	28.08	25.98
D Inds - X Inds	-2.50	-4.20	-6.85	-8.45	-3.20	-4.23	-5.31	-5.86	-16.90	-12.79	-9.07	-7.44	-6.29
Hv Chem Inds	49.20	43.43	41.25	38.52	41.59	42.53	41.60	37.07	39.67	40.86	32.81	31.08	27.72
Light Inds	42.30	43.02	49.05	41.96	40.32	40.04	35.94	36.79	37.11	33.89	32.13	30.41	28.96
Light - Hv Chem	-6.91	-0.42	7.79	3.44	-1.27	-2.48	-5.66	-0.28	-2.56	-6.96	-0.68	-0.67	1.25

/a The figures are the ratios of total bank loans and foreign loans over total asset of each sector.

Source: Financial Statement Analysis, BOK, various issues.

Table 5.8: DEBT RATIO OF EACH SECTOR

	1969	1970	1971	1972	1973	1974	1975	1976	1977	1978	1979	1980	1981	1982	1983	1984
Tot Mnf	270.00	328.40	394.20	313.40	272.70	316.00	339.50	364.60	367.20	366.80	377.10	487.90	451.50	385.80	360.30	342.70
L Mnf	272.90	331.80	402.10	319.10	276.50	322.40	351.50	372.20	369.30	363.30	377.50	504.70	451.10	377.60	360.60	339.50
S Mnf	106.80	206.80	161.30	138.80	148.80	140.90	185.90	272.20	347.50	388.10	374.30	394.90	453.20	433.00	357.80	366.00
S Mnf - L Mnf	-166.10	-125.00	-240.80	-180.30	-127.70	-181.50	-165.60	-100.00	-21.80	24.80	-3.20	-109.80	2.10	55.40	-2.80	26.50
X Inds	NA	NA	NA	403.90	324.10	402.20	337.00	489.01	479.20	519.50	406.40	496.10	427.90	333.80	325.30	286.50
Doms Inds	NA	NA	NA	282.20	250.10	283.00	314.60	320.30	322.40	316.50	366.00	482.40	469.00	425.90	383.20	379.90
Doms - X Inds	NA	NA	NA	-121.70	-74.00	-119.20	-22.40	-168.71	-156.80	-203.00	-40.40	-13.70	41.10	92.10	57.90	93.40
Hv Chem Inds	NA	NA	NA	315.40	259.20	275.10	284.70	324.00	316.30	344.50	331.30	460.00	407.00	327.90	305.20	307.30
Light Inds	NA	NA	NA	311.70	285.30	365.00	408.70	417.00	437.10	403.40	449.10	527.90	533.40	500.10	472.50	421.30
Light - Hv Chem	NA	NA	NA	-3.70	26.10	89.90	124.00	93.00	120.80	58.90	117.80	67.90	126.40	172.20	167.30	114.00

Source: Financial Statement Analysis, BOK, various issues.

Table 5.9: AVERAGE COST OF BORROWING BY EACH SECTOR

	1972	1973	1974	1975	1976	1977	1978	1979	1980	1981	1982	1983	1984
Tot Manfact	12.00	8.60	10.50	11.30	11.90	13.10	12.40	14.40	18.70	18.37	15.97	13.63	14.42
L Manfact	11.98	8.48	10.49	11.19	11.80	11.91	11.91	14.42	18.42	18.30	16.08	13.71	14.45
S Manfact	14.16	11.59	11.41	13.92	14.39	13.80	15.55	14.16	20.74	18.77	15.38	12.95	14.13
S Mnf - L Mnf	2.18	3.11	0.92	2.73	2.59	1.89	3.64	-0.26	2.32	0.47	-0.70	-0.76	-0.32
X Inds	11.06	9.78	9.82	9.82	11.34	12.87	12.68	15.70	16.01	15.81	13.55	12.39	12.91
Doms Inds	12.46	9.84	10.88	12.60	12.25	13.24	12.25	13.80	21.03	20.36	17.59	14.37	15.20
D Inds - X Inds	1.40	0.06	1.06	2.78	0.91	0.37	-0.43	-1.90	5.02	4.55	4.04	1.98	2.29
Hv Chem Inds	10.53	8.65	10.38	10.24	10.14	11.50	10.09	12.51	17.58	17.49	15.29	12.93	14.39
Light Inds	13.31	10.90	10.59	12.16	13.70	14.29	15.85	16.62	20.05	19.64	16.93	14.63	14.46
Light - Hv Chem	2.78	2.25	0.21	1.92	3.56	2.79	5.76	4.11	2.47	2.15	1.64	1.70	0.07

Source: Financial Statement Analysis, BOK, various issues.

Table 5.10: AVERAGE RATE OF RETURN ON INVESTMENT BY EACH SECTOR

	1972	1973	1974	1975	1976	1977	1978	1979	1980	1981	1982	1983	1984
Tot Manfact	10.55	12.77	10.91	9.50	10.42	10.27	11.03	10.74	9.14	9.98	8.80	9.59	9.67
L Manfact	10.55	12.75	10.86	9.43	10.37	10.06	10.82	10.57	8.84	9.70	8.82	9.50	9.59
S Manfact	10.42	13.61	12.81	10.87	11.22	12.25	12.23	11.81	11.10	11.44	8.68	10.30	10.21
S Mnf - L Mnf	-0.13	0.86	1.95	1.44	0.85	2.19	1.41	1.24	2.26	1.74	-0.14	0.80	0.62
X Inds	10.44	15.14	8.21	7.75	9.10	8.75	8.53	8.55	9.22	10.65	7.94	8.91	9.82
Doms Inds	9.23	11.62	12.25	10.82	11.08	11.12	12.33	11.68	9.08	9.54	9.33	9.98	9.58
D Inds - X Inds	-1.21	-3.52	4.04	3.07	1.98	2.37	3.80	3.13	-0.14	-1.11	1.39	1.07	-0.24
Hv Chem Inds	7.92	10.06	12.45	9.34	9.41	8.96	9.69	9.32	7.36	9.11	8.56	9.20	9.75
Light Inds	11.00	15.30	9.45	9.65	11.50	11.57	13.80	12.50	11.40	11.28	9.13	10.15	9.52
Light - Hv Chem	3.08	5.24	-3.00	0.31	2.09	2.61	4.11	3.18	4.04	2.17	0.57	0.95	-0.23

Source: Financial Statement Analysis, BOK, various issues.

Table 5.11: PROFITABILITY OF EACH SECTOR

	1969	1970	1971	1972	1973	1974	1975	1976	1977	1978	1979	1980	1981	1982	1983	1984
Tot Manfact	3.69	2.49	1.00	3.80	7.90	5.70	3.90	4.60	4.50	5.00	3.40	-0.20	0.02	1.03	3.29	3.41
L Manfact	3.64	2.47	0.83	3.75	7.88	5.59	3.78	4.55	4.24	4.93	3.24	-0.59	-0.51	0.88	3.07	3.22
S Manfact	6.60	3.49	4.31	4.93	8.96	9.02	5.96	5.48	5.72	5.30	4.19	1.98	2.68	1.78	4.86	4.65
S Mnf - L Mnf	2.96	1.02	3.48	1.18	1.08	3.43	2.18	0.93	1.48	0.37	0.95	2.57	3.19	0.90	1.79	1.43
				69-75					76-79					81-84		
T Mnf Avg				4.07					4.38					1.94		
L Mnf Avg				3.99					4.24					1.67		
S Mnf Avg				6.18					5.17					3.49		
X Inds	NA	NA	NA	4.55	10.03	2.92	2.45	3.25	2.61	1.78	0.35	0.15	1.30	1.03	3.07	4.33
Doms Inds	NA	NA	NA	2.93	6.87	7.03	4.96	5.28	5.72	6.65	4.67	-0.05	-0.84	1.03	3.38	2.91
Doms - X Inds	NA	NA	NA	-1.62	-3.16	4.11	2.51	2.03	3.11	4.87	4.32	-0.65	-2.14	0.00	0.31	-1.42
				72-75					76-79					81-84		
X Inds Avg	NA	NA	NA	4.99					2.00					2.43		
Doms Inds Avg	NA	NA	NA	5.45					5.58					1.62		
Hv Chem Inds	NA	NA	NA	2.27	5.79	7.63	4.64	4.65	4.55	4.09	3.00	-1.29	-0.26	1.13	3.32	3.71
Light Inds	NA	NA	NA	4.40	9.88	3.80	3.17	4.56	4.51	6.26	3.83	1.12	0.43	0.89	3.20	2.88
Light - Hv Chem	NA	NA	NA	2.13	4.09	-3.83	-1.47	-0.09	-0.04	2.17	0.83	2.41	0.69	-0.24	-0.12	-0.83
				72-75					76-79					81-84		
Hv Chem Inds Avg				5.08					4.07					1.98		
Light Inds Avg				5.31					4.79					1.85		

Source: Financial Statement Analysis, BOK, various issues.

Table 5.12: NET SALES GROWTH RATE OF EACH SECTOR /a

	1969	1970	1971	1972	1973	1974	1975	1976	1977	1978	1979	1980	1981	1982	1983	1984
WPI (All Prod.)	17.10	18.70	20.30	23.10	24.70	35.10	44.40	49.80	54.30	60.60	72.00	100.00	120.00	126.00	126.30	127.20
WPI Grth Rate	1.07	1.09	1.09	1.14	1.07	1.42	1.26	1.12	1.09	1.12	1.19	1.39	1.20	1.05	1.00	1.01
Tot Manfact	NA	NA	15.15	16.88	48.70	10.48	11.47	23.93	21.06	22.76	10.26	-1.36	11.67	6.67	17.72	17.17
L Manfact	NA	NA	NA	17.76	49.64	11.19	12.26	23.93	21.06	21.86	10.26	1.52	11.67	7.62	16.72	16.17
S Manfact	NA	NA	NA	8.09	55.25	3.44	5.14	21.25	21.06	22.76	7.73	-14.32	11.67	5.71	21.71	20.14
S Mnf - L Mnf	NA	NA	NA	-9.67	5.61	-7.74	-7.11	-2.67	0.00	0.90	-2.53	-15.84	0.00	-1.90	4.99	3.97
					72-75				76-79					81-84		
T Mnf Avg					21.88				19.50					13.30		
L Mnf Avg					22.71				19.28					13.04		
S Mnf Avg					17.98				18.20					14.81		
X Inds	NA	NA	NA	NA	58.99	-6.41	7.51	33.73	21.06	20.07	0.16	0.08	15.00	6.67	12.73	25.11
Doms Inds	NA	NA	NA	NA	44.02	21.74	13.84	18.58	21.06	23.65	15.31	-2.08	9.17	7.62	19.71	13.19
Doms - X Inds	NA	NA	NA	NA	-14.96	28.15	6.32	-15.16	0.00	3.58	15.15	-2.16	-5.83	0.95	6.98	-11.92
					73-75				76-79					81-84		
X Inds Avg	NA	NA	NA		20.03				18.76					14.88		
Doms Inds Avg	NA	NA	NA		26.53				19.65					12.42		
Hv Chem Inds	NA	NA	NA	NA	58.05	35.11	9.89	23.93	22.90	25.45	11.94	-2.08	13.33	8.57	19.71	18.16
Light Inds	NA	NA	NA	NA	44.96	-5.00	13.05	23.04	19.23	20.07	7.73	-0.64	9.17	5.71	14.73	16.17
Light - Hv Chem	NA	NA	NA	NA	-13.09	-40.11	3.16	-0.89	-3.67	-5.38	-4.21	1.44	-4.17	-2.86	-4.99	-1.99
					73-75				76-79					81-84		
Hv Chem Inds Avg					34.35				21.05					14.94		
Light Inds Avg					17.67				17.52					11.44		

/a The figures are adjusted for inflation by the formula, 1 + net sales growth rate/1 + WPI growth rate.

Source: Financial Statement Analysis, BOK, various issues.

Table 5.13: ACCESS TO BORROWING BY EACH INDUSTRY
(%)

	1972	1973	1974	1975	1976	1977	1978	1979	1980	1981	1982	1983	1984
Fishing (1)	NA	51.08	46.24	40.14	47.20	54.30	48.49	55.46	42.68	41.47	40.49	31.02	36.75
Mining (2)	39.56	33.05	27.67	26.59	22.88	25.25	21.46	23.78	14.36	12.54	11.08	10.80	7.41
Manufacturing (3)	45.41	43.21	45.22	40.27	40.97	41.32	39.29	36.94	38.55	38.05	32.53	34.73	28.17
Food & beverages (31)	42.44	45.63	43.90	37.18	37.67	29.24	23.18	24.04	25.46	20.25	20.51	16.54	19.59
(Beverage)	40.93	37.39	38.92	34.55	30.21	21.95	26.56	17.89	17.19	16.77	14.51	9.99	12.63
Textile & apparels (32)	47.04	44.59	54.52	44.23	42.24	43.69	42.64	45.14	44.56	39.26	37.50	37.18	34.35
Wood & furniture (33)	47.52	47.76	67.10	59.44	54.24	54.20	44.54	44.33	46.51	51.63	48.46	52.81	52.53
Paper & printing (34)	40.01	34.21	46.77	38.32	39.23	36.73	37.46	32.45	32.31	35.67	27.46	27.61	28.10
Chem, petr. & coal (35)	39.57	39.12	35.15	31.86	31.77	15.11	29.41	28.97	27.01	33.88	28.90	28.95	21.95
Nonmetallic min prod. (36)	67.27	53.57	51.09	44.79	47.35	42.96	40.97	37.61	36.29	35.29	29.80	23.51	22.36
Basic metal (37)	46.37	44.64	43.85	47.14	48.61	49.41	50.64	48.28	57.42	52.55	36.53	32.89	28.84
(Iron & steel)	46.89	45.17	42.02	47.40	48.85	50.52	51.54	51.43	62.38	50.65	36.27	33.16	25.74
Fab met prod & equip (38)	43.87	41.32	40.28	36.64	40.47	38.96	41.56	34.55	37.07	38.19	33.70	31.66	30.35
(Transport equipment)	51.48	47.59	50.01	51.79	51.51	46.54	47.71	34.11	36.56	44.20	34.32	33.13	30.25
Other manufacturing (39)	26.83	29.11	28.41	14.65	31.04	40.48	14.22	32.75	37.03	33.43	29.21	24.43	26.99
Electricity & gas (4)	55.68	49.78	48.77	52.64	27.26	47.58	48.15	53.18	60.70	52.73	50.54	50.48	49.65
Construction (5)	40.90	37.15	44.88	29.26	23.62	20.57	12.67	16.49	29.15	32.68	32.85	26.95	29.56
W & R trade & hotel (6)	27.62	23.01	28.96	20.56	29.52	30.87	28.09	26.08	26.78	23.65	27.04	17.84	21.36
Transport & storage (7)	51.10	44.16	38.08	36.67	38.42	37.03	34.82	34.90	44.06	42.35	31.68	35.36	31.62
Real est & bus serv (8)	30.59	29.95	20.10	15.03	29.54	17.91	30.74	28.29	30.81	11.17	15.88	12.50	9.82
Rec, cul & other serv (9)	31.18	33.76	22.68	17.67	17.02	29.28	21.48	17.46	9.91	12.36	9.04	14.08	18.04

Source: Financial Statement Analysis, BOK, various issues.

Table 5.14: DEBT RATIO BY EACH INDUSTRY

	1972	1973	1974	1975	1976	1977	1978	1979	1980	1981	1982	1983	1984
Fishing (1)	NA	10,361.10	2,160.60	7,839.60	7,281.20	2,361.40	2,850.40	-	2,690.70	1,929.60	3,590.10	1,072.20	2,078.80
Mining (2)	201.40	159.00	153.30	143.00	167.30	240.90	257.40	369.60	262.00	295.00	225.90	181.90	137.20
Manufacturing (3)	313.40	272.70	316.00	339.50	364.60	367.20	366.80	377.10	487.90	451.50	385.80	360.30	342.70
Food & beverages (31)	334.90	375.50	415.60	410.20	453.50	390.50	395.50	409.50	471.90	528.80	422.40	395.60	374.50
Textile & apparels (32)	386.90	310.90	440.60	505.70	497.30	528.10	548.60	638.80	820.10	637.70	598.60	579.60	524.80
Wood & furniture (33)	548.00	442.30	1,688.60	1,751.10	631.10	558.30	560.40	604.70	2,051.40	4,894.80	10,253.00	-2,676.10	10,253.00
Paper & printing (34)	202.10	160.30	243.20	257.50	283.20	310.00	307.50	287.50	382.80	483.80	539.40	393.50	337.20
Chem, petr. & coal (35)	212.80	226.40	257.90	259.30	242.50	276.00	228.20	299.80	414.70	439.20	367.70	297.00	273.30
Nonmetallic min prod. (36)	597.00	373.00	376.00	282.90	323.60	270.30	406.90	422.80	447.90	399.00	290.70	274.40	250.00
Basic metal (37)	552.00	233.80	192.10	282.30	342.20	318.90	301.90	324.90	496.60	463.60	243.80	236.90	219.50
Fab met prod & equip (38)	272.00	245.00	291.90	286.90	381.00	379.00	477.70	335.50	403.00	349.60	374.70	375.70	385.70
Other manufacturing (39)	181.60	167.10	139.10	233.40	262.70	457.90	408.00	598.90	310.00	247.20	355.70	252.70	322.60
Electricity & gas (4)	289.60	330.80	320.60	200.60	133.50	138.80	167.40	197.20	257.90	262.50	158.30	176.90	176.50
Construction (5)	312.40	193.90	306.50	256.60	368.70	300.90	349.80	383.50	524.90	522.80	497.20	481.10	450.80
W & R trade, hotels (6)	190.50	174.30	202.60	281.60	396.10	367.10	431.00	442.10	566.30	590.10	523.20	403.10	465.70
Transport & storage (7)	463.70	342.10	233.00	602.50	419.00	477.70	488.50	449.00	628.40	544.70	485.30	496.90	537.10
Real est & bus serv (8)	176.80	332.70	161.00	416.00	446.00	426.50	185.00	189.00	209.20	183.30	182.10	162.10	285.20
Rec, cul & other serv (9)	9.34	13.05	13.41	99.80	106.80	174.10	141.40	311.00	300.40	252.90	275.50	235.70	196.30

Source: Financial Statement Analysis, BOK, various issues.

Table 5.15: AVERAGE COST OF BORROWING BY EACH INDUSTRY

	1972	1973	1974	1975	1976	1977	1978	1979	1980	1981	1982	1983	1984
Fishing (1)	NA	8.96	7.31	8.17	8.72	8.98	10.64	11.83	14.81	15.90	14.55	13.36	14.44
Mining (2)	17.48	7.98	11.10	8.20	11.00	8.80	8.70	10.80	14.70	17.40	15.00	12.70	14.12
Manufacturing (3)	12.00	8.60	10.50	11.30	11.90	13.10	12.40	14.40	18.70	18.37	15.97	13.63	14.42
Food & beverages (31)	14.58	8.10	10.62	13.90	16.53	16.57	16.80	16.28	20.58	19.36	18.86	16.97	17.04
(Beverage)	(18.85)	(12.59)	(12.87)	(15.76)	(19.66)	(20.95)	(19.89)	(16.52)	(22.60)	(20.04)	(20.23)	(16.03)	(18.51)
Textile & apparels (32)	10.99	8.72	9.05	10.25	12.23	13.54	15.53	16.02	18.92	18.72	15.98	14.55	14.16
Wood & furniture (33)	12.29	6.72	9.80	12.72	9.88	9.83	10.57	13.96	22.39	20.17	15.18	13.26	12.97
Paper & printing (34)	16.80	11.65	14.44	14.99	16.59	17.08	16.96	18.76	20.99	20.09	19.17	13.92	12.21
Chem, petr. & coal (35)	10.33	9.26	13.23	13.11	13.34	12.34	15.30	17.31	23.58	20.85	18.12	13.91	17.45
Nonmetallic min prod. (36)	9.84	10.39	8.33	11.54	10.37	11.34	10.56	13.34	15.72	17.54	16.06	14.69	13.59
(Porcelain)	(26.50)	(14.62)	(11.67)	(11.17)	(18.51)	(18.23)	(12.48)	(15.74)	(29.87)	(24.92)	(19.51)	(14.91)	(14.56)
Basic metal (37)	11.38	5.55	8.09	8.19	7.75	7.73	6.89	9.97	12.24	15.62	11.67	11.36	12.29
(Steel)	(11.81)	(5.93)	(7.93)	(7.77)	(7.85)	(7.28)	(6.63)	(9.67)	(11.36)	(15.00)	(11.23)	(10.52)	(12.47)
Fab met prod & equip (38)	14.57	8.62	12.11	10.61	11.66	12.32	11.51	13.46	19.71	17.51	15.69	13.08	13.98
(Shipbuilding & auto)	(13.88)	(7.27)	(10.50)	(8.87)	(9.28)	(9.27)	(9.28)	(13.02)	(17.35)	(15.58)	(13.15)	(11.50)	(14.15)
Other manufacturing (39)	16.93	11.74	14.60	15.96	14.06	14.86	14.19	17.61	20.33	15.60	15.44	13.04	14.52
Electricity & gas (4)	4.98	4.85	5.30	5.40	5.10	4.00	4.80	5.30	7.40	8.20	5.80	5.70	6.45
Construction (5)	12.69	12.28	11.70	12.90	15.00	11.80	15.30	17.20	18.40	19.50	16.10	15.40	13.55
W & R trade, hotels (6)	17.02	11.18	12.00	14.70	15.80	16.40	16.70	20.30	23.70	24.80	18.40	16.30	17.19
Transport & storage (7)	8.84	5.95	8.30	10.00	8.10	11.90	11.10	12.60	15.60	18.00	16.10	12.90	13.89
Real est & bus serv (8)	17.73	10.49	8.20	7.70	8.80	8.70	4.00	4.70	7.00	16.20	11.40	9.50	11.06
Rec, cul & other serv (9)	16.32	9.03	10.00	8.20	14.10	12.10	11.90	21.10	30.80	18.60	14.00	14.10	10.83

Source: Financial Statement Analysis, BOK, various issues.

Table 5.16: AVERAGE RATE OF RETURN ON INVESTMENT BY EACH INDUSTRY

	1972	1973	1974	1975	1976	1977	1978	1979	1980	1981	1982	1983	1984
Fishing (1)	NA	7.08	1.23	1.43	5.36	5.37	7.13	5.34	2.29	6.17	-1.29	6.30	6.44
Mining (2)	4.36	10.21	16.89	12.03	10.70	6.34	3.81	0.16	3.38	5.36	5.53	3.82	3.88
Manufacturing (3)	10.55	12.77	10.91	9.50	10.42	10.27	11.03	10.74	9.14	9.98	8.80	9.59	9.67
Food & beverages (31)	12.13	11.51	9.48	11.28	13.21	15.36	16.83	13.55	11.97	11.52	10.55	10.97	8.88
Textile & apparels (32)	10.02	16.06	7.42	6.71	9.58	8.30	11.02	10.37	9.82	11.54	8.30	8.59	8.23
Wood & furniture (33)	14.56	16.72	-1.00	7.77	7.71	10.62	14.72	8.24	-0.16	5.04	3.28	8.89	4.67
Paper & printing (34)	13.69	16.04	14.89	11.05	13.75	14.47	14.22	14.53	10.80	7.38	8.71	10.00	9.97
Chem, petr. & coal (35)	10.54	12.60	12.45	12.89	13.16	10.53	14.12	13.98	12.18	10.91	10.75	10.98	12.40
Nonmetallic min prod. (36)	12.72	18.75	15.77	12.26	10.15	12.81	10.53	11.96	10.50	9.39	9.56	11.66	12.16
Basic metal (37)	7.83	8.95	18.18	4.12	7.09	7.56	7.70	8.38	6.24	9.56	6.61	8.35	9.62
Fab met prod & equip (38)	10.64	10.92	12.26	11.70	10.03	11.25	8.78	8.93	7.04	8.71	8.45	9.04	8.81
Other manufacturing (39)	11.12	11.77	12.89	17.32	13.67	11.40	9.15	6.93	12.52	14.34	12.95	13.54	12.07
Electricity & gas (4)	4.98	4.51	4.12	5.09	5.48	7.60	6.61	11.28	10.91	9.00	6.50	6.31	8.36
Construction (5)	11.34	10.86	10.72	12.87	15.49	18.41	17.14	13.36	12.24	11.70	10.49	9.74	7.79
W & R trade, hotels (6)	8.53	10.11	11.20	12.42	13.05	11.30	11.85	11.95	11.48	11.19	8.57	9.19	8.29
Transport & storage (7)	7.38	9.77	6.93	7.28	6.58	8.94	8.60	9.39	8.33	9.94	8.27	4.73	5.53
Real est & bus serv (8)	9.42	9.35	9.24	6.98	6.57	7.81	8.32	8.57	7.67	11.58	8.66	9.36	5.58
Rec, cul & other serv (9)	131.20	132.20	90.50	12.91	21.10	15.03	19.30	16.30	16.65	7.62	8.17	10.95	10.36

Source: Financial Statement Analysis, BOK, various issues.

Table 5.17: PROFIT RATIO OF EACH INDUSTRY

	1972	1973	1974	1975	1976	1977	1978	1979	1980	1981	1982	1983	1984
Fishing (1)	NA	0.68	-2.66	-3.61	0.28	-0.65	0.53	-1.61	-6.28	-1.49	-8.73	-0.26	-0.55
Mining (2)	-0.29	6.58	12.80	8.90	7.10	8.20	1.10	-3.90	-0.60	0.80	1.63	1.00	0.82
Manufacturing (3)	3.80	7.90	5.70	3.90	4.60	4.50	5.00	3.40	-0.20	0.02	1.03	3.29	3.41
Food & beverages (31)	4.06	6.36	4.10	4.09	5.10	8.23	10.46	6.10	2.96	10.67	2.86	4.39	2.20
Textile & apparels (32)	3.47	10.61	2.10	0.92	3.20	1.40	2.93	1.15	-0.71	0.53	-0.05	1.29	1.65
Wood & furniture (33)	7.58	12.16	-7.00	-0.48	1.32	4.52	9.01	0.95	-13.10	-9.24	-6.08	-1.23	-4.70
Paper & printing (34)	5.15	10.51	7.55	3.68	5.05	6.11	6.12	5.47	-0.17	-3.72	0.08	3.55	3.96
Chem, petr. & coal (35)	5.59	8.17	7.03	7.70	7.90	5.21	8.42	6.84	3.29	0.81	2.57	5.08	5.59
Nonmetallic min prod. (36)	0.40	3.63	3.59	5.79	4.46	6.75	4.84	4.71	2.06	-0.56	1.01	5.03	6.35
Basic metal (37)	1.41	5.65	14.35	3.20	3.15	3.42	3.75	2.66	-1.09	-0.05	0.73	3.19	4.66
Fab met prod & equip (38)	2.04	6.17	6.86	6.84	4.41	5.58	2.82	1.86	-2.94	-0.30	0.68	2.74	2.43
Other manufacturing (39)	5.48	6.69	8.29	10.48	6.70	4.95	2.74	-0.47	3.27	7.47	7.39	9.09	6.21
Electricity & gas (4)	1.54	1.09	0.80	2.00	2.80	5.60	4.10	8.40	6.40	4.40	3.10	3.20	4.74
Construction (5)	4.59	5.54	5.20	7.90	10.50	14.90	13.50	8.90	5.40	3.00	2.70	2.90	1.40
W & R trade & hotel (6)	2.58	6.54	7.00	8.10	7.50	5.40	5.90	3.80	2.40	2.10	1.50	2.70	1.70
Transport & storage (7)	2.17	5.84	3.00	2.00	1.80	3.90	3.40	3.20	0.60	0.80	1.10	-1.20	-0.24
Real est & bus serv (8)	3.61	4.43	7.20	5.00	3.30	6.00	7.00	6.90	5.30	9.00	6.30	7.30	3.73
Rec, cul & other serv (9)	4.43	8.89	10.30	11.10	18.10	12.20	16.10	10.40	10.60	3.80	6.00	8.20	7.39

Source: Financial Statement Analysis, BOK, various issues.

Table 5.18: NET SALES GROWTH RATE OF EACH INDUSTRY

	1972	1973	1974	1975	1976	1977	1978	1979	1980	1981	1982	1983	1984
Fishing (1)	NA	63.05	-11.38	14.69	40.01	9.65	3.33	-11.24	-16.36	8.27	-9.84	14.56	9.32
Mining & quarrying (2)	-2.99	19.34	15.13	3.72	9.66	19.50	2.95	10.09	-6.90	17.75	4.57	5.95	8.48
Manufacturing (3)	17.14	48.98	10.62	11.47	23.57	20.69	22.31	9.84	-1.22	11.58	6.95	17.52	16.73
Food & beverages (31)	11.83	32.06	3.86	16.44	3.87	18.54	18.46	13.97	-2.39	5.97	1.51	21.55	10.39
Textile & apparels (32)	34.29	57.94	-10.31	19.81	32.82	18.60	19.72	-0.33	4.72	14.67	4.46	6.54	18.74
Wood & furniture (33)	15.75	45.64	-29.49	-1.56	28.56	18.11	18.21	7.50	-17.18	-4.23	4.68	23.19	2.14
Paper & printing (34)	8.41	33.96	-1.13	-7.29	20.78	18.66	8.42	13.64	2.47	12.87	14.52	26.75	16.09
Chem, petr. & coal (35)	11.06	39.58	41.51	19.75	14.65	14.07	14.20	17.58	13.34	8.72	3.90	11.31	9.19
Nonmetallic min prod. (36)	7.43	30.27	14.39	31.41	15.93	22.80	10.07	15.72	-4.44	1.41	11.17	23.82	12.20
Basic metal (37)	30.95	105.35	30.20	-16.32	35.46	20.27	28.99	14.00	-4.10	12.68	9.47	19.76	12.05
Fab met prod & equip (38)	10.24	69.94	16.15	7.98	37.76	35.17	40.53	6.71	-18.01	18.56	13.05	30.38	29.28
Other manufacturing (39)	9.96	37.06	-2.60	11.37	18.09	33.45	18.41	8.12	-20.55	7.25	10.87	11.31	32.51
Elec, gas & water (4)	13.35	14.48	16.32	51.47	18.76	19.78	10.66	37.86	19.74	14.08	10.57	8.74	8.70
Construction (5)	-9.99	15.59	-5.56	15.58	51.92	96.08	72.58	25.41	3.97	22.92	17.71	7.34	-0.39
W & R trd & rest & htls (6)	52.53	38.65	-6.48	24.83	54.42	40.78	39.87	18.00	9.94	13.25	10.67	18.12	18.27
Trans, storage & commun (7)	16.53	48.87	3.87	13.76	11.45	30.78	25.18	16.91	2.53	7.42	6.57	6.75	6.10
Fin, ins, R est & bus srv (8)	-6.65	54.31	21.25	7.75	50.32	65.27	5.91	44.09	-13.89	20.92	19.43	27.40	20.81
Social & personal servs (9)	-18.47	15.96	17.59	18.11	26.78	34.27	24.46	7.90	-10.94	-1.25	15.43	7.44	22.96

Source: Financial Statement Analysis, BOK, various issues.

<u>Table 5.19</u>: KOREA

	1969	1970	1971	1972	1973	1974	1975	1976	1977	1978	1979	1980	1981	1982	1983	1984
Average cost of borrowing	13.39	14.66	13.40	12.00	8.60	10.50	11.30	11.90	13.10	12.40	14.40	18.70	18.37	15.97	13.63	14.42
WDI growth rate	6.88	9.36	8.56	13.79	6.93	42.11	26.50	12.16	9.04	11.60	18.81	38.89	20.00	5.00	0.24	0.71
GNP deflator	14.80		13.92	16.11	13.40	29.54	25.73	20.73	15.67	21.89	21.16	25.63	15.90	7.08	2.90	3.90
Average cost of borrowing - WDI growth rate	6.51	5.30	4.84	-1.79	1.67	-31.61	-15.20	-0.26	4.06	0.80	-4.41	-20.19	-1.63	10.97	13.39	13.71
Normal profit ratio	3.69	2.49	1.00	3.80	7.90	5.70	3.90	4.60	4.50	5.00	3.40	-0.20	0.02	1.03	3.29	3.41
Net earnings ratio			4.50	16.70	30.30	22.70	16.50	21.60	21.30	22.90	15.60	-1.30	0.10	5.30	15.50	15.20
Debt ratio	270.00	328.40	394.20	313.40	272.70	316.00	339.50	364.60	367.20	366.80	377.10	487.90	451.50	385.80	360.30	342.70

Source: <u>FSA</u>, BOK, various issues.

Table 5.20: JAPAN

	1969	1970	1971	1972	1973	1974	1975	1976	1977	1978	1979	1980	1981	1982	1983	1984
Average cost of borrowing	8.50	8.60	8.60	8.10	8.40	10.30	9.60	9.40	8.50	7.30	7.70	9.60	9.40	8.40	8.00	NA
WDI growth rate	2.20	3.60	-0.80	0.80	15.70	31.60	3.00	5.00	1.90	-2.60	7.30	17.80	1.40	1.80	-2.20	NA
Average cost of borrowing - WDI growth rate	6.30	5.00	9.40	7.30	-7.30	-21.30	6.60	4.40	6.60	9.90	0.40	-8.20	8.00	6.60	10.20	NA
Normal profit ratio	6.90	5.90	3.90	4.70	7.10	4.70	1.40	3.20	3.40	4.00	5.80	5.30	4.30	3.90	4.10	NA
Net earnings ratio		29.20	19.90	24.30	38.30	26.20	8.30	18.90	19.50	22.60	30.10	26.40	20.80	17.60	17.40	
Debt ratio	NA	403.00	421.00	424.00	449.00	459.00	488.00	488.00	475.00	446.00	418.00	385.00	378.00	342.00	324.00	NA

Source: FSA, BOK, various issues.

Table 5.21: BANK'S CREDIT ALLOCATION BY INDUSTRY
(%)

	1973	1974	1975	1976	1977	1978	1979	1980	1981	1982	1983	1984	1985
Fishing (1)	9.43	6.71	7.14	10.05	10.04	6.71	5.63	5.45	7.77	5.04	11.29	11.91	0.7
Mining & quarrying (2)	-1.91	1.01	1.10	2.14	-0.06	1.14	1.56	0.34	-0.36	1.43	0.04	-0.42	0.6
Manufacturing (3)	68.22	65.26	52.94	60.50	59.41	49.96	59.58	59.94	58.75	37.01	36.10	35.83	45.9
	(19.20)	(20.60)	(21.60)	(23.40)	(24.10)	(26.60)	(27.60)	(28.80)	(29.10)	(28.60)	(28.90)	(30.90)	
Elec, gas & water (4)	2.10	3.31	9.71	3.87	5.06	5.64	4.37	3.35	4.27	1.60	2.55	2.67	4.4
	(1.20)	(1.20)	(1.40)	(1.40)	(1.50)	(1.60)	(1.70)	(2.00)	(2.00)	(2.10)	(2.20)	(2.20)	
Construction (5)	6.68	10.51	2.50	6.92	6.75	10.27	15.06	12.01	2.14	13.47	23.40	27.60	24.1
	(6.80)	(6.50)	(6.90)	(6.70)	(7.50)	(8.40)	(8.10)	(8.40)	(7.50)	(8.50)	(9.50)	(8.80)	
W & R trd & rest & htls (6)	3.98	4.79	7.02	8.56	7.21	12.86	4.22	6.20	8.35	17.01	8.75	9.66	10.4
Trans, storage & commun (7)	4.76	3.23	14.08	6.41	8.90	6.72	7.25	8.03	9.19	11.44	8.12	4.87	11.2
	(5.40)	(5.30)	(5.50)	(5.40)	(6.00)	(6.50)	(7.10)	(7.70)	(7.80)	(7.80)	(7.70)	(7.50)	
Fin, ins, R est & bus srv (8)	3.55	1.94	3.42	0.89	0.77	2.77	1.45	1.75	3.64	0.72	6.65	3.00	1.6
Social & personal servs (9)	3.18	3.24	2.08	0.65	1.93	3.93	0.88	2.93	6.25	12.29	3.11	4.89	1.1

Note: The figures are the share of net credit increase of DMBs and KDB. The figures in parenthesis are share of value added.

CHAPTER 6: SHIPPING: A CASE STUDY OF INDUSTRIAL RESTRUCTURING

A. Introduction

6.01 The shipping industry is a valuable case study of industrial policy gone awry. For a number of reasons, including some public externalities related to national security but also based on the view that shipping services could add to the value-added of exports and reduce the value-added of imports, shipping was encouraged by government incentives. The ensuing expansion of capacity was financed with generally small amounts of equity and many firms entered the market. Average annual growth in Korean shipping capacity was 21.8% over the 1962-83 period. Unfortunately the industry has suffered net losses in recent years and a rationalization program was recently imposed by government to bring about a reduction in the number of firms and to provide the framework for a rollover of some of the industry's debts. Although government intervention in the merchant fleets of maritime nations is common, it runs counter to the Korean government's current policy intentions and thus merits more detailed analysis as a lesson for future actions. The following section describes the basic facts of the industry.

Basics of the Industry

6.02 Shipping is a capital-intensive industry providing a generally homogeneous service, although there are various lines of service depending on the material to be shipped. It is convenient to distinguish between dry bulk carriers, tankers, and container ships. Shipping is a worldwide market, with a number of very large firms and a host of smaller ones. Korea's share of gross world tonnage was approximately 1.5% (as of 1983) with slightly higher shares of the dry bulk, dry/liquid (i.e., convertible) bulk, and general cargo trade and considerably less of the tanker and container trade. According to KMPA, the size of Korea's merchant fleet was some 917 ships as of 1983, accounting for just over 6 million gross tons (g.t.) compared with a mere 114,000 g.t. in 1962 and 1.03 million g.t. in 1972. Thus, the average annual growth in shipping capacity was a healthy 23.2% in the 1962-1972 period and 18.3% in the subsequent 1973-83 period.

6.03 Compared with its contemporaries, Korea's shipping industry has grown somewhat more slowly than Singapore's and marginally less than Hong Kong's over the past decade, although its fleet ranks second in current gross tonnage. Korean ships have in part by virtue of reservation-waiver systems [1/] received increasing shares of Korean export and import cargoes: export shares rose from roughly 38% (1972-77) to 46% (1978-83) and import shares climbed

---

1/ The reservation-waiver system refers to the requirement that certain national bulk cargoes be carried by Korean-flag vessels unless waivers are requested and approved. Included in these commodities are imports of iron ore, coal, raw chemical products, grain, fertilizer, crude oil, and government purchases as well as exports of steel, cement, and plywood. The system of cargo preference is legislated in the Shipping Promotion Act of 1967.

from 32% (1972-77) to 49% (1978-83). Overall capacity utilization rates have exceeded world averages, even in difficult years like 1983;[2] however, the industry has not been profitable in 4 of the past 5 years being analyzed, and because of its heavy indebtedness has required government-sponsored relief efforts.

6.04 Korea's most recent surge in capacity, an increase of 62% between 1979 and 1983 has coincided with constrained world shipping demand. Therefore while some additions to capacity have taken place in container ships, and to a lesser extent dry bulk and other ships, global capacity reductions have been effected in tankers, dry/liquid bulk carriers, and conventional general cargo vessels. Somewhat surprisingly, in at least two of these categories Korea had added significantly to its capacity in recent years (see Table 6.1). By virtue of primarily second-hand purchases, Korea boosted its capacity in all types of ships except tankers. Given the general slowdown in world trade growth, and the fact that about 40% of total Korean tonnage is in cross-trade,[3] the bullish expectations of Korean shippers must have been fueled either by (i) high hopes for Korean-origin or destination trade; (ii) lower anticipated costs of shipping enabling Korea to take new market shares; or (iii) very effective public incentives for expansion, perhaps based on non-economic strategic objectives.

6.05 It is important not to underestimate the strategic factor, as greater shipping self-sufficiency was a Korean national objective. While it is true that a certain amount of economic miscalculation occurred (viz., the somewhat unanticipated global slowdown), Korean shippers adopted a risky strategy of purchasing many fuel-inefficient and manpower-intensive ships in hopes of operating them at lower costs. The Korean urgency in buying second vessels created what was called the "Korean price," essentially highly inflated prices. This strategy, whether correct or not, could never have proceeded without Government's explicit support in the form of financing, although it should be noted that foreign banks also undertook this type of financing.[4]

6.06 Korea's shipping trade grew rapidly in the 1972-83 period of examination in concert with Korea's overall trade expansion. Export and import tonnage increased six- or seven-fold during the period and Korean flag vessels' share also increased markedly. Shipping accounted for about 26,000 jobs in 1982 (5.7% of total transportation employment) as well as some indirect employment in shipbuilding, although the latter impact is modest with only

---

2/ In dry bulk shipping, for example, 81% of Korean capacity was used in 1983 compared to a 55% worldwide average and in tanker traffic, 67.7% of Korean capacity was used versus a 41% average worldwide.

3/ Total tonnage in 1983 was accounted for by exports (14%), imports (46%), and cross trade (41%).

4/ Of course, foreign banks consider business loans to be essentially sovereign risk.

Table 6.1: COMPARATIVE SHIPPING CAPACITY
(millions gross tons)

	Bulk carriers			Tankers	Container ships	Other	Total
	Dry	Dry/liq	Gen'l cargo				
World 1979	81.80	26.50	80.20	183.20	10.00	31.30	413.00
World 1983	98.40	26.00	78.30	169.80	14.10	35.90	422.60
Change	20%	-2%	-2%	-7%	42%	15%	2%
Korea 1979	1.23	0.03	0.81	1.31	0.20	0.37	3.95
Korea 1983	2.87	0.71	0.99	1.00	0.31	0.51	6.39/a
Change	133%	2,374%	22%	-23%	50%	38%	62%
Japan 1979	9.50	3.30	4.40	18.00	1.40	3.50	40.00
Japan 1983	11.50	2.00	3.70	17.50	1.70	4.20	40.80
Change	21%	-38%	-16%	-3%	25%	27%	2%

/a Current estimates put total tonnage at slightly above 7 million gross tons.

Source: Lloyds Register of Shipping as reported in "An Overview of Prospects and Strategy for the Development of the Korean Shipping Industry," Report for the IBRD by Temple, Barker & Sloane and the Korea Maritime Institute (July, 1985).

14.8% of gross tonnage (and 12.4% of value-added) produced in Korean shipyards being for domestic firms. Shippers have relied primarily on imported purchases (79% of total tonnage between 1977 and 1983), much of it second-hand. Still, the average age of Korea's fleet at 10.9 years is equivalent to the world average, but older than either Hong Kong (7.4 years), Singapore (9.6 years) or Japan (7.9 years). It is reported that almost 25% of Korean ships are 15 years old or more. In terms of average ship size, Korea's fleet is considerably smaller (3,685 g.t.) than either Hong Kong (14,910 g.t.) or Singapore (8,198 g.t.) but not too different from Japan's (3,847 g.t.). Still Japan's ore and bulk carriers and container ships are considerably larger than Korea's, although Korea's fleet is increasing in terms of average tonnage in recent years.5/

5/ Between 1979 and 1983, dry bulk ships averaged a 31% increase in average tonnage to almost 20,000 g.t. (compared with Japan's 32,000 g.t.) and container ships increased 38% in average weight to almost 11,800 g.t. (compared to 25,700 g.t. in Japan).

6.07 Korea's cost structure, at least with respect to operating costs, was competitive or better than that of its major rivals. And its capital equipment costs were also competitive, inasmuch as second-hand vessels purchased entail lower initial outlays and are often financed with credit from official institutions. The industry has produced about a third of the invisibles account revenue in recent years and about 9% of total foreign exchange earnings. Since no firm data is available on repayments of foreign loans by shippers, we cannot say what the net foreign exchange gain of the industry really is, however.

6.08 To summarize Korea's position in world shipping, it should be noted that Korea (i) expanded its fleet considerably, even in recent years when world demand was slack; (ii) concentrated on buying generally smaller, second-hand vessels because of their ready availability and lower cost; and (iii) successfully expanded its very small market share in most shipping categories (see Table 6.2), but did so at a price, i.e., by generating losses. These losses were difficult to absorb by the many highly leveraged firms in the market and this put Government, which had a role in the promotion of the industry, squarely into the "workout process" which was required when these highly indebted firms got into financial difficulties. The following section provides some background on the economic situation of the industry.

Table 6.2: SHIPPING SHARES
(percentages)

Tankers	Dry bulk	Gen'l cargo	Container	Total
		Shares of World Capacity (1983)		
0.59	2.92	1.27	2.16	1.51
		Share of World Capacity (1979)		
0.72	1.51	1.01	2.04	0.96
		Share of World Demand (1983)		
0.40	2.38	0.65	1.53	"
		Share of World Demand (1979)		
0.38	1.17	0.53	1.28	"

Source: Temple, Barker & Sloane.

The Economics of the Industry

6.09 Korea managed to do well in world shipping on a volume basis. Korean-based trade expanded and in the vital market of cross-trade (non-Korean trade), Korea managed to increase its market share from .21% in 1979 to 1.25% in 1983 at a time when the overall volume of cross-trade was steadily declining.[6/] In order to capture this difficult share--and keep utilization rates considerably above world averages--Korean shippers lost money. As noted in Table 6.3, the average return on sales (1979-83) was -1.1%, the average return on assets was -0.9%, and the average return on equity was -6.8%. Given an average debt-equity ratio of 8 to 1 as of 1983 and a rate of return of -29% in that year, it was clear equity was going to be eroded to the point where debts were nonrepayable. Firms lost a combined W 158 billion in 1984, pushing total debts to W 2,828 billion at the end of 1984, of which W 1,232 billion was short-term debt.

Table 6.3: THE FINANCIAL SITUATION OF THE SHIPPING INDUSTRY

	Return on sales	Return on assets	Return on equity	Debt-equity ratio
	(percentage)			
1979	-0.2	-0.1	- 1.4	8.6
1980	-1.6	-1.4	-13.8	8.9
1981	0.4	0.3	2.2	5.6
1982	-2.9	-2.2	-14.0	5.4
1983	-4.1	-3.2	-28.8	7.9
1979-83 avg.	-1.1	-0.9	- 6.8	7.1

Source: Korea Maritime Institute and TBS (1985)

6.10 On a disaggregated basis, there is no component service of Korean shipping that was profitable in 1983, as seen in Table 6.4. In terms of assets, short sea haulers and special carriers account for about 11% of the total, while liner fleets account for 45%, bulk carriers 28%, and Southeast Asian carriers 16%. The biggest problem area is the bulk market which included 41 companies as of 1983 and in which firms were very highly leveraged and in which competition was ruthless. Freight rates below already depressed international prices were often quoted with the objective of survival rather than profit generation.[7/] While this strategy is evident in other fields of Korean business (viz., it has received tacit approval by government through its reward system), Korean shippers were not alone in their desperation to cover some part of costs. Based on capacity utilization comparisons, however,

---

6/ Total world seaborne trade fell by about 20% between 1979 and 1983.

7/ The daily rate quoted for a Panamax class bulk carrier (55,000-65,000 d.w.t.), for example, fell from $14,000 in 1981 to $4,000 in 1983 and often below that.

it can be concluded that they were more successful than others in capturing market shares.

6.11 The resulting structure of the industry as of 1983 was generally as follows: there were 70 operators in the industry, but the 5 largest accounted for 44% of total capacity. Hyundai alone accounted for 13% of fleet capacity. While the number of crude oil carriers increased from 2 operators in 1970 to 4 in 1980, the number of nonliquid bulk carriers exploded from just 10 in 1970 to 35 in 1975, 43 in 1980, and 41 in 1983. Liner carriers increased rather extensively as well from 9 in 1970 to 21 in 1980 and 23 in 1983. Thus, by 1983 there was a plethora of firms in the shipping industry, competing vigorously with one another and trying to gain market shares by outbidding international competitors in a declining revenue market. Completing the dualistic structure of the industry were the 42 operators owning less than 50,000 g.t. of capacity and constituting 16% of total fleet size.

Table 6.4: THE FINANCIAL SITUATION OF MAJOR COMPONENTS OF THE SHIPPING INDUSTRY, 1983

Trade	Sample number of companies	Return on sales	Return on assets	Return on equity	Debt-equity ratio	Proportion of long-term debt
		---------	(percentage)	--------		
Liner /a	4	-1.0	-0.8	-6.2	6.6	60
Bulk/b	31	-9.7	-6.3	-91.7	13.6	59
SE Asia/c	7	-2.8	-2.6	-18.0	5.8	45
Short Sea/d	8	-11.2	-9.8	-71.0	6.3	63
Special/e	4	-4.6	-3.4	-52.3	14.7	50
Total	54	4.1	-3.2	-28.8	7.9	--

/a Liner trade refers to regularly scheduled service, and in the case of Korea, refers to container ships.
/b Trade carried on a contractural or irregular basis (tramp service).
/c Liner trade to S.E. Asia.
/d Liner trade to Japan.
/e Ships designed for automobiles, refrigerated cargoes, chemicals, etc.

Source: Korea Maritime Institute

6.12 This industrial structure was not viable, and probably would not have arisen in the marketplace without explicit incentives because creditors would not normally finance new entrants into an industry not even earning normal profits. By the end of 1984, the shipping companies as a whole were saddled with debts (end 1984) of W 2,824 billion, a slight increase (W 54 million) over 1983. This debt level when compared with equity of W 274 million reveals a current debt-equity ratio in excess of 10. Of the total amount of debt outstanding, W 546 billion was to due Korean banks in 1985, much of it for past purchases of ships, and had to be rescheduled (over 8 years with a 3-

year moratorium) by the Government as a follow-up to its 1984 rationalization-rescue package. To put this set of public interventions into perspective, it is useful to review the public sector's role in the shipping industry.

B. Public Sector Role

Government Philosophy

6.13 Shipping takes on major strategic significance in the Korean setting. Faced with North Korea on one side and water on the other three, Government has made it a priority to be able to support itself satisfactorily by sea. This has in turn meant that public monies have tended to be used to produce shipping services which are perceived to be in the public interest. Inasmuch as shipping firms are privately owned, this philosophy has brought about classic intervention. Intervention has taken the form of legislation which provided specific industrial incentives, ad hoc financial measures aimed at supporting or rescuing ailing firms, and government-led industry restructurings, complete with forced mergers and rescheduling of debts.

6.14 Several interesting features of the shipping industry emerge as a result of Government's industrial policy. Explicitly, the incentive system and the low equity requirements encouraged a large number of entrants. Ultimately this resulted in virulent competition, undercutting of prices, and the sustaining of large losses as the combination of excess supply and poor world demand conditions coexisted. In a way, however, this mirrors industrial behavior in other sectors, where very active competition (often with different economic agents gaining unequal access to resources, for example, as with respect to credit) was encouraged to foster "the survival of the fittest." This was often the first step to further growth, industrial concentration, and further economic rewards. Thus, an implicit set of incentives encouraged firms to overexpand and overborrow in hopes of being one of the ultimate industry survivors and gain access to preferred sources of credit. While the strategy has benefits, it also entails certain costs.

6.15 One disadvantage of this approach, and one which is exacerbated by the weaknesses of the financial sector is that it ultimately brings in Government as industrial arbiter and gives the public sector a role in picking survivors. This is often accomplished, as in the Japanese case, through arranged mergers. In these cases, the larger firms tend to absorb the smaller ones, which validates the grow-at-all-costs strategy and rewards those who grew fastest rather than those which are most efficient or most profitable. Although there are cases where third-party intervention among competitors is useful,[8] public sector decisions are often based on noneconomic factors such as regional economic strength, balance among conglomerates or sheer guesses about survivability of firms, which again is often biased in favor of large firms.

8/ There are similarities to the prisoners' dilemma in same theory among feuding firms in an industry characterized by excess capacity.

Incentive Structure

6.16 The Government has taken an active role in promoting and now rescuing the shipping industry. This support has taken the form of direct and indirect financial support, including direct operating subsidies, preferential access for capital equipment (ship) purchases, tax incentives, subsidized credits for domestically produced ships, and most recently rationalization programs which work hand-in-hand with debt rescheduling. The Shipping Promotion Act of 1967, which incidentally codified the system of cargo preferences for Korean vessels, authorized operating subsidies totalling W 835 million which were disbursed over the 1967-73 period.[9/] Financial support appears to have been ad hoc in order to bolster survival prospects of particular firms. It is not clear to what extent subsidiaries of conglomerates also benefitted from subsidies, but evidence points to their having benefitted.

6.17 In 1973, the subsidy program was discontinued, although a program of grants to the operators of full container vessels was instituted in 1976 with the objective of developing that segment of the industry. The beneficiaries of these grants (four major firms) received government grants averaging W 770 million annually during 1979-1982 and W 325 million annually during 1983-84. It is not clear whether this support was prompted by infant industry arguments, or national security, concerns or other factors. Container capacity did expand between 1979 and 1983 by 50%, but most firms have lost money consistently since 1979.

6.18 Prior to 1982, Korean shipping companies were exempt from corporate income tax. Subsequently, they have been liable to corporate taxation at the basic 30% rate. Exemptions are currently given for transfer and acquisition taxes in connection with the rationalization program which brought about wholesale mergers of shippers. Recent legislation will also exempt shippers and other declining industries from capital gains taxes in the disposing of real estate and other assets needed to pay off overdue bank loans as well as exempt the banks themselves from these taxes in selling of collateralized assets.[10/]

6.19 Most of the buildup of Korea's shipping that has come by way of the second-hand market, but about 12% of new capacity installed during the decade ending in 1983 was domestically built. This figure is higher (i.e., 17%) for the subperiod 1977-82. Therefore, some of the benefits of incentives offered to the domestic shipbuilding industry indirectly accrued to the domestic shipping industry by way of subsidized ships, and others were directly targetted on Korean shippers to help finance the purchases of Korean vessels. These

---

9/ In addition, 394 million won was used to compensate two liner companies for 1970-73 losses incurred. See Korean Shipowners' Association for details.

10/ Tax Reduction and Exemption Law of 1985.

latter subsidies are not that dissimilar from incentives offered by other maritime producers, however.

6.20 The major vehicle for this kind of support in the late 1970s was the Planned Shipbuilding Program of 1976. In 1981, the Shipping Promotion Fund was established and qualified ocean-going carriers were eligible for special domestic financing 11/ in connection with purchases from Korean shipyards. The original objective was to expand shipping activity in new trades and invest W 10 billion annually; however, in actuality, only W 1 billion was allocated to the fund in 1981 and it was discontinued in 1983. Under the recent rationalization program, however, the grace period was doubled for these loans. In 1984 a new program for the conversion of insufficient tonnage was introduced with construction loans of up to W 10 billion being allocated under similar terms.

Recent Industrial Policy Dilemma

6.21 Government was recently faced with the following problem: shipping firms were losing money and a large number of them were unable to service their debts. Bankruptcies were considered generally not in the national interest because: (i) the financial sector was already straining under the difficulties caused by tardy repayments of shipbuilding and overseas construction loans; (ii) shipping capacity was considered to be in the national interest and certain forms of service were essential to maintain; and (iii) such actions could prompt a more generalized loss of confidence in Korea's creditworthiness and imperil its foreign debt exposure and cost of borrowing. The approach selected was a government-sponsored Shipping Industry Rationalization Program (SIRP).

6.22 The rationalization program offers support in the form of tax exemptions, ship conversion funds, and most importantly, loan moratoria to those firms agreeing to be merged within a group. The Korea Maritime and Port Administration (KMPA) proposed the formation of some 17 new groups of carriers in its rationalization program presented to the Industrial Policy Council, chaired by the Deputy Prime Minister (EPB) and including the Ministry of Finance and the Ministry of Trade and Industry. This recommendation was based on work done by the Korea Maritime Institute, which sought to preserve carriers in the deepsea bulk, shortsea bulk, deepsea liner, and shortsea liner routes.

6.23 KMPA's guidelines for newly created companies were as follows:

- key deepsea bulk companies should include at least 8 smaller companies contributing a total of 500,000 g.t. and should command aggregate capacity of at least 1.3 million g.t.;

11/ Financing of 70-75% of cost at 10% over a 10.5 year period (grace period of 2.5 years) was normal.

- key deepsea liner companies should include at least 5 component firms contributing a total of 300,000 g.t. and should exceed 1 million tons in total capacity;

- for shortsea bulk trade, a tramp company for Korea-Japan and 3-4 tramp companies for Korea-South East Asia (each with about 20 ships) is desirable; and

- for shortsea liner trade, one container company for Korea-Japan, 2 breakbulk companies for the same route, and one combined container/breakbulk company for Korea-South East Asia trade is desirable.

6.24 The objectives of these forced mergers has been to (i) reduce cutthroat competition among many small firms, particularly in bulk trade; (ii) achieve certain economies of scale by closing down redundant facilities; (iii) encourage the scrapping of outmoded capacity; (iv) establish efficient service on a number of important routes; and (v) ultimately improve the financial condition of the industry. The specific inducements offered are that firms qualifying under the program will be able to extend the grace period of loans given to buy domestically built ships or second-hand ships from the existing 2.5 years to 5.0 years, and therefore push out the final maturity to 13 years. Banks agreed to these 1984 rollovers, and Government guaranteed up to $300 million (roughly W 255 billion) of these loans.[12] SIRP participants are allowed to take out foreign currency loans to repay both principal and interest. An additional W 3 billion will be made available for the conversion of inefficient tonnage into energy and manpower saving ships. The aim is reported to be a removal of about 300,000 g.t. of aged capacity annually between 1985 and 1987, reducing the fleet tonnage by about 1 million g.t.

6.25 The net result of the SIRP is that 15 shipping groups have been formed out of the existing 63. The remaining 3 firms in the core industry declined to participate in the rationalization program. Firms were encouraged to sell off nonshipping assets in addition to disposing of antiquated tonnage.[13] In this respect, the government has acted in ways similar to traditional creditors in enforcing "workouts." This may not be inappropriate since the KDB is the major domestic creditor for the industry, at least for debts arising out of ship purchases. It is not clear, however, what rules were applied to conglomerate owners of shipping companies (such as Hyundai,

---

12/ Outstanding indebtedness of the industry is reported (Korea Herald, July 21, 1985) to be W 2,828 billion, so the guarantee applies to almost 10% of total debt.

13/ As a result of the rationalization program, about 440,000 tons of capacity has been disposed of, 540 local and overseas branches have been closed, and 900 workers have been laid off. Combined efforts to sell assets resulted in only W 72 billion in new funds compared with losses of W 158 billion in 1984 alone, prompting Government to amend the tax provision as noted in para. 7.6.

Daewoo, and Hanjin) and whether these subsidiaries benefitted from the SIRP bailout pari passu with other firms.

Assessment of Intervention

6.26 There are a number of possible approaches for assessing interventions. One can ask how effective the action was in bringing about the desired objective. Or one can compare the ultimate outcome to the counterfactual case of nonintervention. Or one might propose alternative interventions which might produce similar or clearly superior (normative) outcomes. Taken one by one, the measures introduced served to achieve the short-run objectives of policymakers, i.e., bailing out struggling firms, forcing mergers in an industry with too many small firms, prodding some scrapping of obsolete or redundant capacity (reportedly 385,000 g.t. in 1985), and maintaining a vital public service. But this was brought about at a substantial cost.

6.27 The direct cost includes implicit subsidization of banking institutions, primarily the official KDB which holds a large share of the total outstanding debt of the shipping industry, but also privately owned commercial banks to permit them to rollover debts of as much as W 546 billion in 1985. In the near term the question is whether the SIRP reorganization will produce sufficient efficiency gains to produce profits in 1986 and 1987.[14/] Firms are being encouraged to divest themselves of other assets, but these often are insufficient to cover the losses of smaller or nondiversified firms.[15/] Moreover, divestiture of aging vessels will produce far less than is required to repay the respective loans on these ships. Thus, the Government has (i) essentially invested permanently in this troubled industry and (ii) turned much of the industry over to conglomerates.

6.28 Government's actions were clearly aimed not only at bailing out firms, and implicitly lenders, but also in assuring continued shipping services of various kinds. This concern with the public goods aspect of shipping led it to explicitly design the industry's postbailout structure. The approach is not too dissimilar from the Japanese approach to industrial restructuring, where Government and industry design industrial units using efficiency and financial criteria as well as criteria relating to geographic employment and industrial needs. The normative question is how good government is at designing industrial structure and how different that managed is structure from unimpeded market structure.

6.29 The first observation is that in the presence of neutral incentives, firms should themselves see the merits of agglomeration, if they exist, and

---

14/ Debt service due is reported to be at least W 352 billion in 1986 and W 288 billion in 1987.

15/ One large firm, for example, disposed of W 20 billion in real estate and securities but this does not even match the W 24 billion won in current year losses incurred.

the industry itself should move to capture the economies of scale and positive rents. Why did this not happen in Korean shipping, an industry in which many new firms entered over the past 5 years and most lost money? In an industry with large capital requirements and economies of scale, there are normally financial barriers to entry. In the Korean case, many new firms were allowed to borrow with low equity at relatively inexpensive rates. Shippers could borrow for Korean or imported ships at approximately 10% over 10 years, rates presumably not reflective of the actual credit risks involved. In addition, special loans to shipowners were authorized to assist in debt service repayments between 1977 and 1979 and again in 1982 for foreign currency loan repayments.[16/] Government also provided shippers with short-term working capital loans, subject to conditions,[17/] equal to W 225 billion in 1983, compared with roughly W 200 billion in 1982 and W 140 billion in 1981.

6.30 Some conclusions to be drawn from the existence and scope of these support programs are that they were pervasive, somewhat unpredictable or on-off in nature, and apparently targetted on the specific problems of the moment. As such, the incentive structure was not consistent or far-sighted enough. A second conclusion is that the SIRP and the relief programs that preceded it neither got to the roots of the problem nor forced firms to stand on their own two feet. Therefore, one might expect additional relief measures to be needed. This prognosis relates to a third conclusion, which is that the efforts made to avoid bankruptcy of firms may serve to produce a suboptimal industry structure.

6.31 In container shipping, for example, the top 30 operators worldwide each have a fleet capacity in excess of 10,000 TEUs (20-foot equivalent units). A legitimate issue might relate to the appropriate economic size of the Korean container fleet and the process by which industrial size is determined. Over time, to be competitive internationally it is likely that Korean container companies will have to achieve economies of scale in feeder-relay traffic and intermodal transport. This should and can be accomplished by the private sector, but will probably mean even less competition. Merely for comparison, it should be noted that total Korean capacity combined would only place Korea sixteenth internationally in fleet size. All of the top 30 operators have fleet sizes in excess of 11,000 TEUs and Korea's largest company (KSC) has about 10,000 TEUs in capacity. Moreover, given current technology, viable international container-ship firms need to achieve economies of scale in both land and ocean sides of transport, particularly since

---

16/ Loans to help shippers repay mortgages totaled 8.2 billion won over the 1977-79 period and W 172.5 billion over the 1982-83 period. Loan terms were 3 years (1 year grace), interest rate of 16%, maximum of 50% of total debt service in any year, limited to certain types of shipping services which excluded liners in Japan and South East Asia trade as well as tankers, and extendable after 1983 if shippers are SIRP participants.

17/ The terms of these 90-day loans were 10% and the loan amount was subject to 10% of annual freight revenue.

higher profit margins will require investments in the upstream and downstream aspects of shipping, such as terminals and land transport.

6.32 In bulk shipping, it is widely recognized that the incentive structure encouraged the entry of too many firms and that capacity needed to be reduced. Voluntary mergers (and scrapping) or bankruptcies (and forced sales) are very viable alternatives to managed restructurings, in particular since it is more likely in both cases that the more efficient, and ultimately more profitable firms, might survive. Forced mergers involving a neutral organizer can be justified in a market characterized by predatory competition if it is deemed in the national interest to prevent bankruptcies. This apparently was the public policy decision in Korea. The acid test is whether the newly created groups can regain profitability and ultimately repay their loans.[18/] The second important issue relates to Government's future actions with respect to these rescheduled debts if they cannot be repaid. And a related and equally important issue concerns what future government policy towards the industry will be, particularly in light of KMPA's announced plan to see Korean shipping capacity more than doubled by the year 2000. These forward-looking issues are discussed in C below.

C. Industry Outlook, Strategy, and Issues for Industrial Policy

Industry Outlook

6.33 Forecasting medium-term supply is far easier than forecasting demand inasmuch as shipbuilding has a relatively long gestation period and demand factors are a function of global growth performance. Still, forecasting of both supply of and demand for shipping services is done and does provide some clues as to the appropriate balance between the two.[19/] On the basis of global shipping projections, one can reach some very tentative conclusions concerning the outlook for the Korean industry. Since more than half of Korea's capacity is in dry bulk carriers, this is the most important market segment. It is predicted that the demand/supply ratios will continue to fall in the mid 1980s from their 1982 levels, especially for smaller ships (which

---

18/ Looking to the future, there is a projected oversupply in the 10,000-80,000 g.t. segment of bulk shipping, which is particularly acute for Korea. The largest remaining bulk shipping company, Pan Ocean, with roughly 20-25% of the Korean fleet, absorbed losses in 1984 almost equal to its equity and is reported to have a debt-equity ratio of 20 to 1. Business Korea reports Pan Ocean liabilities of W 700 billion (August 1985) and equity of W 35 billion (October 1985).

19/ See, for example, Temple, Barker & Sloane, 1985.

characterizes the largest part of Korea's bulk fleet).[20/] While Korea is quite competitive with other fleets with respect to vessel costs, these amount for only 20% of total costs, as capital expenses are about 40% and voyage costs a similar percentage.

6.34 It is not at all clear that Korea will increase its share of the all important cross-trade market. Indeed, the recent bankruptcy of Sanko, the largest Japanese shipping firm, which began a program of acquiring 125 new bulk carriers in 1984, casts a dark shadow over the industry in East Asia. Financial problems have spread throughout the shipping industry; for example, Hong Kong shipping firms of considerable magnitude are seeking restructuring agreements from their creditors.[21/] It is also reflected in the record pace of tanker scrapping, which reported totalled 26.6 million tons in 1983, 20 million tons in 1984, and 30 million tons as of November 1985.[22/]

6.35 One of the important issues to be addressed is whether or not the maintenance of the current share of Korean and cross-trade cargo by Korean bulk carriers is a national priority. If it is, then the normal accretion of trade volumes and the rather high rate of scrapping which would occur between 1984 and 1990 (in part due to the high average age of the fleet) would necessitate an investment of $2.3 billion to rebuild obsolete tonnage.[23/] If second-hand tonnage is purchased, the replacements would cost less than half that much. In order to remain price competitive, however, shippers would have to receive substantial financial assistance in the purchase of ships, adding to the burden being placed on the government and perpetuating the debt servicing difficulties of the industry.

6.36 In the technologically more advanced container market, in which Korean ships command 30% of Korean export trade (1979-83 average) and 38% of Korean import trade, estimates for the 1983-1990 period indicate that overall demand may increase by about 7.3% annually.[24/] Since 1978, transpacific container rates have fallen dramatically [25/] and given the known new supply coming on-stream in the next few years, capacity utilization rates are expected to fall. Since the largest container companies (i.e., US and those

---

20/ Of 7.66 million dead-weight tons (DWT) of dry bulk capacity in 1983, only 2.40 (31%) was larger than 80,000 DWT. DWT is the usual measure of capacity and includes maximum cargo plus stores, while gross tonnage refers to the vessel's weight.

21/ See Far Eastern Economic Review (December 5, 1985).

22/ Much of this involves the very large crude carriers (200,000-300,000 tons) or the ultra large crude carriers (over 300,000 tons).

23/ Korea Maritime Institute and TBS (1985).

24/ Ibid.

25/ From an indices of 100 to 70 (Eastbound) and 53 (Westbound), as of 1984.

of a nearby Asian island economy) have invested in fuel-saving ships and are achieving new economies of scale by use of round-the-world (relay system) routes, it is unlikely that Korean shippers will be gaining market shares. Indeed, inasmuch as land transport feeder systems and marketing factors favor the large firms,[26/] Korea will have to strive to maintain its existing share. It is estimated that investments of between $300-500 million are needed to maintain a 30-40% share of the Korean container trade by 1990.[27/]

6.37 It does not appear that shipping is an industry in which Korea has a natural comparative advantage, nor can one be easily developed. There are no large cost differentials to exploit and decisions will have to be made to either accept smaller market shares (and presumably risk continued non-payment of debt obligations) or invest once again to try and maintain market shares. Investment in second-hand ships will perhaps make it harder to compete technologically and with respect to price, while the alternative of investing in new ships is very costly, risky, and difficult for an over-indebted industry to handle. In the immediate future, the reorganized shipping groups must reduce redundant capacity because a continuation of the past "ship-at-any-price strategy" is not viable. Over the 1979-83 period, for example, Korean bulk carriers were losing money (negative returns as sales), while the industry average was positive. The same picture existed in container shipping. It is fairly self-evident that losses must be taken on some of these assets vis-a-vis their book values and some debts will go unpaid. In light of these factors, it is difficult to separate the industry's strategic planning from public policy, since the public sector is the major creditor of shipping firms and because Government continues to see significant public merit in the industry's preservation.

Strategy

6.38 By all accounts, Korea is not inclined to abandon its shipping industry. It has placed restraints on its medium-term expansion, however, sanctioning "new bottoms" (capacity increases) only if they are matched by assured future cargoes. A case in point is automobile-carrying ships, which will be needed to move Korean cars to North America. In a departure from past practices, Korean shipyards are to be given preference in new ship orders, and the Planned Shipbuilding Program invisions about W 200 billion in support for 1986.[28/] Based on the evidence of 1985 in terms of new tonnage added (for the transport of cars and steel), it appears that the retrenchment will focus on

---

26/ Although there are almost 200 container operators, the top 30 (each with 10,000 TEU capacity or more) account for 75% of total volume shipped.

27/ Korea Maritime Institute and TBS (1985).

28/ This would be financed by W 160 billion from the National Investment Fund and W 40 billion from the KDB. Borrowing costs are now 11.5% for NIF loans and 13.0% for KDB loans. Financing is available for up to 90% of ship price. Preference is given to firms in the SIRP and those with guaranteed cargoes.

carrying Korean traded commodities and reconsider cross-trade secondarily (if and when world demand recovers) but on a long-term reservation basis if possible. New tonnage in 1984 was 826,000 g.t., which is certainly not indicative of an industry in rapid decline. Since scrapping estimates are 385,000 g.t., Korea's fleet actually expanded by 441,000 g.t. during 1985.

6.39 The second aspect of Korea's future strategy which appears quite clearly is the increasing reliance on large firms. This means that the industry will not only be more concentrated, but also that it will in large measure involve the conglomerates. In container traffic, in particular, there will be only three majors, now that Hyundai has taken over a heavily indebted Korea Maritime Transport Company; the other being Hanjin and Korea Shipping Corporation. In noncontainer business, wholesale mergers have also taken place, either managed by Government as part of the SIRP on privately (viz., Pan Ocean and Daeyang recently merged to form Korea's largest shipping line).

Public Policy

6.40 Shipping poses an industrial policy problem for Government. On the one hand, an examination of the operating subsidies, preferential credit, and bank rollovers shows an industry highly dependent on public resources. This kind of resource drain is contrary to government's expressed desire to wean industries from perpetual support and force them to survive in a more competitive market environment. On the other hand, because of the extent of Government's direct intervention, to the point of dictating the industry's organizational structure, it now continues to have a stake in the industry's success. At some point, however, the viability of separate components of the industry must be assessed, the realization reached that sunk costs are not retrievable, and a judgment reached as whether further investments of public monies are warranted. At the moment, the bulk carrier market is being bolstered on the demand side by the reservation-waiver system and on the supply side by financial assistance. There is a legitimate issue concerning the desirability of maintaining these props over time. The former might be justified on national security grounds, and it serves to guarantee a certain minimum level of demand for bulk cargoes. The latter is difficult to justify over time and does not serve to promote the infusion of new equity which is needed.

6.41 Some of the steps which would be desirable include: (i) a clear timetable for the withdrawal of direct financial support to the industry; (ii) debt rescheduling conditionality including mandatory profit plowbacks, divestiture, and a halt to further highly leveraged ship purchases; (iii) the tying of any new tax incentives to infusions of new capital, particularly from parent firms; (iv) a clear signal that bankruptcies will be tolerated; and (v) writeoffs of nonrepayable government loans to prevent the continued lending of resources to nonprofitable enterprises. Similar to the case of Sanko in Japan, banks will have to absorb losses. Unlike Japan, however, the banking sector in Korea relies heavily, and in the first instance, on Government for its economic health.

6.42 In a way, shipping provides clear lessons for the conduct of industrial policy in other sectors of the economy. First, past government interventions should not lead to perpetuation of costly actions. At some point, all creditors cease investing new funds. Second, the merits of discontinuing the management of inefficient units of economic activity should be considered. In the near term, losses and dislocations will inevitably occur, but in the long term, economic efficiency will be enhanced by market-based rather than administratively convenient solutions. Third, the financial sector, which is on a path towards greater managerial autonomy, needs to be extricated in some fashion from the errors of the past, and, must objectively consider the costs and risk-adjusted benefits of its investments. Shipping is a good case where Government's role needs to be more clearly delineated and reduced.

CHAPTER 7: KOREAN TEXTILES: CASE STUDY OF AN INDUSTRY IN TRANSITION

A. Introduction

7.01 The production of textiles and clothing in different countries has tended to follow the product cycle sequence associated typically, but not exclusively, with the growth of income and wages. A country begins this sequence as an importer of a good; it then moves to an import-substitution stage as its labor force acquires the skills and technology to produce the good domestically; the third stage is one of exports as the country, having moved up on the technological learning curve for the product, begins to take advantage of its relatively lower production cost. Through these stages the standard of living rises and wage costs rise correspondingly. As these costs rise beyond a certain point, the country begins to lose its comparative advantage in the product and must yield its export market share to other countries which are a stage behind in the product cycle sequence. Competition in foreign markets is only the first volley of a barrage that typically lays siege to the domestic market also. It is at this point, when export markets are slipping and the domestic market is threatened, that a hard public policy choice has to be made between protecting the industry from competition or allowing international market forces to restructure the industry with the ensuing displacement of labor and capital. If the latter course is chosen, the country moves on to the fourth stage where it becomes an importer of the product once again.

7.02 Korea's textile industry is somewhere in the third stage now and finds itself facing a struggle to maintain export shares in some major products and markets. Competitive forces within the country have already begun to allocate resources away from the sector. That a difficult part of the third stage has arrived can be judged from the following recent developments. Korea's share of the imports of clothing in the United States, by far its most important market, has been virtually stagnant for the past five years. The rate of growth of textile output has slowed to a crawl of only 5% (during 1979-84) from an average of 20% between 1970 and 1978 (see Table 7.1). Capacity has declined or remained stationary in large segments of the industry. The ratio of profits to assets has slipped from an average of 2% (during 1973-77) to less than 1.0% (during 1979-84). The industry's share in manufacturing value added, employment and exports has declined steadily over the past five years. The index of revealed comparative advantage (RCA) of Korea's major manufactured products indicates that textile and clothing products have suffered the biggest relative decline since 1973.

7.03 At this point it is useful to note three caveats. First, the industry is considerably heterogeneous with respect to product and process characteristics and different subsectors have evolved differently and face different prospects. In particular, the clothing subsector must be distinguished from the others. Second, the product-cycle concept is more of a heuristic device than a predictive one. Comparative advantage in a product is affected by a host of factors and it is very difficult to distinguish between cyclical and structural changes in competitiveness. The former are transitory and may be influenced by macroeconomic policy whereas the latter are permanent and denote the onset of major changes in output and employment structure. Third, the industry may be in relative decline but it is not yet in absolute decline. Output and exports continue to rise although employment appears to have peaked in 1982.

Table 7.1: REAL GROWTH RATES OF THE TEXTILE INDUSTRY, MANUFACTURING SECTOR AND GNP, 1968-84
(%)

Year	GNP	Manufacturing sector	Textile industry
1968	11.2	27.2	22.5
1969	13.7	21.6	20.3
1970	7.6	19.9	21.6
1971	8.8	18.7	25.7
1972	5.7	14.7	28.6
1973	14.7	29.9	38.9
1974	7.5	15.7	9.3
1975	6.9	11.9	18.9
1976	14.1	23.9	22.5
1977	12.7	16.0	4.7
1978	9.7	21.3	13.1
1979	6.5	10.3	3.2
1980	-5.2	-1.1	5.1
1981	6.2	7.2	14.7
1982	5.6	4.0	-1.1
1983	9.5	10.9	5.8
1984	7.6	14.6	4.0
1970-78 (A)	9.9	18.9	19.8
1979-84 (A)	4.9	7.5	5.2

Notes: Growth rates are based on value-added in 1980 prices.
A = Average annual growth rates

Source: Economic Planning Board.

7.04 These caveats notwithstanding there is an increasing sense of malaise about the industry in Korea. It is often openly referred to as a sunset industry and public policy initiatives to manage a "soft landing" for this erstwhile high flier are being actively debated. This paper reviews the policy options that Korea faces with respect to the future evolution of the textile industry.[1/] This review is based partly on an assessment of the factors that have influenced the past development of the industry and those that govern its future prospects and partly on an assessment of the experience of developed countries in dealing with the problems of their textile industries over the past two decades. The paper is organized as follows. Section A provides a quick overview of the industry and its role in Korea's economy. Section B examines the factors that have influenced the recent evolution of the industry. Sections C and D review the private and public sector aspects of the adjustment strategy that seems to be emerging. Section E describes the main elements of the adjustment strategies followed in Japan, West Germany, Italy, the UK and the US. Section F discusses the future prospects of the Korean industry. Section G summarizes the lessons for public policy that one may derive from the preceding sections.

B. Overview

7.05 The textile industry's share in manufacturing value added and employment has declined steadily from 16.7% and 26.5% (in 1981) to 13.8% and 22% (in 1984), and its share of total exports has slipped from 29.9% (in 1979) to 24.2% (in 1984). Despite its recent decline the industry remains an important force in Korean manufacturing and exports (see Table A7.1). In 1984 it contributed almost a seventh of value added in manufacturing and close to one fourth of total exports ($7 billion out of $29 billion). It also provided employment for 738,000 persons out of a manufacturing work force of 3.3 million, making it the largest single sector for nonagricultural employment in the country. The industry also has significant linkages with other sectors, especially the machinery and chemical sectors (42% of textile machinery used in 1980 was of domestic origin).

7.06 The textile industry in Korea consists of four major subsectors distinguished by process and output: spinning (yarns); weaving and knitting (fabrics); apparel (clothing and other made-up goods); and artificial or chemical fibers. All these subsectors grew rapidly during the 1970s (the industry grew at an annual rate of almost 20% in real value added terms between 1970 and 1978) and virtually all have experienced a sharp slowdown in recent years (the industry has grown at only about 6% during 1979-84). Variations in recent growth performance, however, can be seen by examining the behavior of the physical output at the subsector level (see Table A7.2). During 1979-84, the production of textile yarns has grown between 5.2% (cotton yarn) and 10% (woolen yarn), that of fabrics has ranged from 8.0% (cotton) to a decline of 16.6% (rayon), and that of chemical fibers from 19.1% (polyester) to a decline of 10% (viscose). The relatively slow growth of the industry as a whole is an

---

1/ Unless otherwise specified, the textile industry will denote both textiles and clothing in this report. Detailed statistics pertaining to output, capacity, employment and other indicators are provided in the appendix.

average of the fairly high growth of some products and sharp declines in the output of others.

7.07 Developments in production capacity suggest a similar story. Over the period 1979-84, capacity has remained virtually stationary in spinning, has declined in knitting and weaving but has grown fairly rapidly in sewing (garment-making) and producing chemical fibers (see Table A7.3). All this reinforces the view that while some causes of the recent slowdown may apply to the entire industry a more productive line of inquiry would focus on subsector-specific causes and issues.

7.08 The intensification, in recent years, of the problems besetting the industry has had a noticeable effect on its profitability. The ratio of profit to assets for the industry has slipped from an average of 3.6% (during 1973-77) to about 1.0% (during 1978-84).

7.09 The industry has always been export driven in that exports have always accounted for a significantly larger fraction of output than domestic consumption. In recent years (see Table A7.5) almost 62% of the output of fibers, yarns, fabrics and clothing has been exported. One measure of its export orientation is the fact that Korea is now the third largest textile exporter in the world (behind Italy and West Germany and ahead of Japan).

7.10 There is considerable variation in economic characteristics across subsectors. For example, over 50% of the employment in the industry is accounted for by the apparel subsector (see Table A7.6). This subsector is characterized by a large number of small, often family-owned, firms who work as subcontractors to large General Trading Companies and operate with relatively simple technology and limited cash. The average firm employs about 40 workers (see Table A7.7).

7.11 At the other extreme are spinning, weaving and knitting mills. These tend to be large-scale and generally capital-intensive operations. The average mill size in cotton spinning in Korea is about 150,000 spindles.[2/] The artificial fibers subsector is also characterized by a small number of large-scale firms using advanced technology and employing few people per unit of capital.

2/ This is larger than the average cotton spinning mill in the US and in Europe. In terms of employment, spinning, weaving and knitting mills together account for about 35% of total employment in textiles. As could be expected these subsectors have few firms. There are only 23 cotton spinning mills in Korea (as of 1983) and about 90 woollen and worsted mills. The average cotton spinning mill employs about 2,300 workers and the average woollen mill about 650 workers.

C. Factors Affecting Recent Evolution of Industry

Protectionism

7.12 An important global development of recent years has been the upsurge of protectionism in developed countries (DCs), especially directed against textile and clothing imports from developing countries. The growth of such protectionism can be seen in the nature of the various multilateral agreements to restrict imports which go under the name of the Multifiber Arrangement (MFA). Since the early 1970s such agreements have become increasingly restrictive and there is no prospect of more liberal agreements in the near future. This upsurge in protectionism can be traced back to the generally sluggish performance of the OECD economies in the 1970s. Slow growth has meant little room for job creation and relocation of displaced workers. This has both increased pressure from industry associations and trade unions in these countries and softened government resistance to protectionist arrangements. Textiles remain an important, though declining, component of the manufacturing sectors in most OECD countries and textile lobbies enjoy considerable political power and influence.

7.13 The rise in protectionism has affected Korea's exports in the past but will probably be of more significance in the future.[3/] Textile exports grew at a rapid pace of 43.5% per year (compounded nominal rate) during 1972-78 but have since slackened, in the face of world recession and protectionism, to an average of under 10% p.a. (simple nominal rate) during 1979-84 (see Table A7.8). Successive MFA's have widened the range of products covered and reduced the scope for quota growth for many of Korea's textile products. Bilateral arrangements have also been forced upon Korea by several of its major trading partners to further restrict the scope for export and output growth.

Aging Machinery and Technical Developments

7.14 There are two noteworthy aspects related to machine vintage in the Korean textile context. The first is that a large and rising proportion of existing stock is old by engineering standards. While only 31% of spinning machines in operation were over 10 years old in 1980, almost 45% were in this category by 1983. Similarly, the proportion of looms over 10 years old rose

---

3/ A review of Korea's quota utilization rates indicates that, as a rule, quotas are not fully utilized (see GATT, 1984, pp. 92-96) Does this indicate that such bilateral agreements do not actually restrict trade? There are several reasons to avoid making such an inference. There typically are mismatches between demand and supply; rapid changes in fashion may make some quotas, based on historical demand patterns, redundant while others are quickly filled. The average quota utilization rate may not capture this phenomenon adequately. Redundancy may also occur when a recession in the importing country leads to a sharp reduction in demand (as in 1981-82). Therefore, an underutilized quota need not indicate that the restriction does not affect trade.

from 31% to 51% during 1980-83 and that of (false) twisters from 33% to 65% (see Tables A7.9 and A7.10). The second is that significant technical innovation has taken place in the art of spinning and weaving during the past two decades in several OECD countries. The confluence of these two trends with rising domestic wage costs has jeopardized Korean competitiveness in certain textile items. The threat from the technology side is greatest in spinning and weaving operations and least in garment-making.[4/]

7.15 These developments have a number of implications for Korea: (i) the labor cost advantage, especially in spinning and weaving, is no longer as great as it used to be; (ii) developed countries can divide textile processes into two activities: the spinning and weaving can be done at home with the new machines while the garment-making can be done abroad in LDCs, thereby cutting costs and also tying in Korean competitors to lucrative OECD markets; and also (iii) while Korea can take advantage of the availability of new machines this does threaten the linkage between domestic machinery suppliers and the textile industry.

7.16 The increase in the age of spinning and weaving machines is surprising because the average factory is large and ought therefore not to have any problems with financing. Perhaps the problem here is not the availability of finance but "diseconomies of scale" arising from largeness of size. What these mills may need is not just better machines but a change in product mix together with a change in the production technology. They may not have yet mastered the necessary concomitant changes in management, design and marketing which would make new machines a worthwhile investment.

7.17 Even the apparel industry where Korea used to enjoy a great cost advantage, has come under pressure from cost and quality concerns. Labor costs have risen sharply in Korea and rapid changes in tastes and fashions have also occurred. Total orders per piece have dropped substantially (from 20,000 to less than 1,000 per design according to one estimate) thereby placing companies with high volume machines at a disadvantage. Given present quality demands, it is estimated that over 30% of the sewing machines in the industry's stock are outmoded.

4/ There have been some dramatic technological developments in the industry during the past 20 years or so. These developments include: open-ended rotor spindles which lead to a 40% saving in manpower requirements over traditional ring spindles in spinning operations; shuttle-less multiphase looms in weaving that significantly reduce labor requirements, breakages and weaving time; high-speed knitting machines which can be patterned via computer controls and offer both reduction in time and increase in flexibility; special purpose machines which perform operations as button holing, patch pocket stitching, decorative stitching, cloth cutting and high speed sewing--all these have made clothing manufacture much less labor intensive than previously; and computer controlled design facilities which have increased the speed and range of design innovations.

Competition from China and Others

7.18 After a long period of self-imposed exclusion from world trade in textiles China, the world's largest textile producer, has begun to make its presence felt in world trade. China was not among the top 15 clothing exporters in 1963 and was the 15th in 1973; ten years later it had risen to 6th place. China possesses two advantages over Korea: (i) it has much lower labor costs and can oust Korean from markets for low quality textile goods and garments; and (ii) it has not yet been subjected to very restrictive quotas since it is a new participant. While the latter source of advantage will undoubtedly be reduced by protectionism in OECD markets, the former source will continue to threaten Korea's position. Furthermore, if world trade in textiles slows or is restricted greatly by protectionism, giving China a bigger slice than it has had historically will mean giving already established exporters like Korea a smaller slice.

7.19 China's share of the Japanese import market doubled in the last decade from 8.4% in 1971 to 16.8% by 1980. Its penetration of the US market has been even more spectacular (see Table 7.2). In 1973, Korea's share of

Table 7.2: COMPETITION FROM CHINA IN THE US MARKET /a
(%)

	1963		1973		1978		1982	
	T	C	T	C	T	C	T	C
Japan	23.7	26.5	19.1	11.5	21.3	5.1	19.4	3.1
UK	8.3	5.9	6.6	2.3	5.7	1.4	4.1	-
Italy	7.1	24.3	6.3	5.5	9.1	3.2	8.4	2.5
Two major Asian producers /b	3.2	17.7	7.3	37.0	8.8	41.9	10.2	43.0
Korea	-	1.0	1.3	11.3	2.7	17.3	6.5	17.3
China	-	-	-	-	2.9	1.1	8.6	7.8

/a Share of selected suppliers in import market for textiles and clothing in US, 1963-82.
T = textiles; C = clothing.

/b Includes Hong Kong.

Source: GATT, 1984, Tables 2.19, 2.20

the US import market for textiles was about 1.3% while China's was non-existent. By 1978 both Korea and China possessed about 2.8% of the market. By 1982 China had taken a clear lead as its share rose to 8.6% while Korea's rose to 6.5%. This experience was repeated in an even more striking fashion in the US import market for clothing. While Korea's share stagnated at 17.3% over 1978-82, China's grew sharply from 1.1% to 7.8% (Hong Kong's share also stagnated in 1978-82). Just as East Asia ousted Japan from the US market for clothing, so does China now threaten the East Asia group.5/

7.20 Another threat on the horizon is India whose potential power in exports is similar to China. India's strength lies in clothing exports, hitherto of ethnic specialties but, in the future, also of mill made fabrics and clothes. The reason India will be a force to reckon with in the future is because of recent changes in its industrial and trade policies which should make its products competitive in export markets.

7.21 One adjustment measure that many textile companies in developed countries have taken progressively through the 1970s is relocation of some activities to low-cost areas. This has resulted in a large amount of outward processing trade (OPT). In particular, German firms have made increasing use of East European countries for OPT, French firms of the Mediterranean rim countries, and American firms of ASEAN countries and Central America. Mexico, in particular, has long provided a location for American garments manufacturing. Export processing zones in Malaysia, Sri Lanka and Mauritius have also attracted DC investors in significant numbers. Korea faces competition from many of these countries to the extent that its labor costs are higher than theirs, its machinery not superior to that installed by DC foreign investors in these "overseas" operations, and its location not as favorable to serve many major consumer markets.

Declining Credit Availability

7.22 The Korean manufacturing sector has been built on commercial bank debt more than on any other source of finance. Debt to equity ratios tend to be high in Korea relative to the three other major Asian producers (including Hong Kong and Singapore). The textile industry was built in the sixties through easy access to credit provided through the government-run banking system. Not only was the industry favored in obtaining finance for investment and working capital it was further supported at the marketing stage by export promotion loans made available to export activities. In the mid-seventies, however, the Government shifted priorities away from light industry to heavy industry in such a massive way that the textile industry was starved of funds. Declining profitability since the late seventies has aggravated the credit access problem. Loans and discounts of commercial banks to the textile industry have declined since the mid 1970s (see Table 7.3). Loans and discounts of the SMIB to this sector, a major source of credit for the bulk of the industry's firms, have also declined in similar fashion (see Table 7.4). The cost of capital for the textiles sector has gone from an average of 9% during the early 1970s to around 15% over 1980-84. It used to be lower than the average cost of capital in manufacturing in the early seventies but has generally been higher since 1978 (see Table 7.5). This has come about because textile firms have been forced to borrow more and more from the curb market

---

5/ The areas in which China is competitive are: silk yarn and fabrics, wool sweaters, cotton fabrics, garments, work gloves, etc. It has also begun entering the synthetic fiber/products market.

where interest rates are much higher, on average, than in the formal banking sector.

Table 7.3: SHARES OF CREDIT ALLOCATION BY BANKS /a
(%)

	Textiles and clothing	Heavy and chemical industry	Light industry
1973	42	36	64
1974	35	32	68
1975	13	66	34
1976	35	56	44
1977	27	61	39
1978	23	56	44
1979	19	59	41
1980	17	60	40
1981	20	53	47
1982	11	69	31
1983	14	59	41
1984	26	56	44

/a Figures refer to the percentage share of the net credit increase of deposit money banks and the Korea Development Bank.

Source: Bank of Korea, Economic Statistics Yearbook.

Table 7.4: OUTSTANDING LOANS AND DISCOUNTS OF THE SMALL AND MEDIUM INDUSTRY BANK, /a
(Million won)

As of end of period	All industries (A)	Growth rate (%)	Textile, wearing apparel and leather (B)	Growth rate (%)	B/A
1975	190,673		57,695		30
1976	230,612	21	76,212	32	33
1977	300,754	30	91,812	20	31
1978	426,968	42	129,015	41	30
1979	645,331	51	168,926	31	26
1980	970,538	50	239,838	42	25
1981	1,434,365	48	328,367	37	23
1982	1,759,426	23	408,624	24	23
1977-82 /b		40.7		32.5	

/a Small and medium industries are industries employing less than 300 employees.
/b Average annual growth rate.

Source: Small and Medium Industry Bank.

Table 7.5: COST OF BORROWING /a

	1970	1975	1978	1980	1982	1983	1984
Textiles	9.4	9.6	15.4	17.4	16.9	14.6	13.6
Clothing	6.7	17.5	15.1	21.0	9.8	10.6	14.6
Manufacturing	14.6	11.3	12.4	18.7	16.0	13.6	14.4

/a Cost of borrowing is measured by ratio of financial expenses to total borrowing.

Source: Bank of Korea, Financial Statement Analysis.

7.23 One consequence of declining profitability, reduced access to credit and increased costs of credit has been a sharply rising debt equity ratio in textiles. This ratio rose from 5.06 in 1976 to 21.55 in 1982 in the clothing subsector and from 4.52 in 1977 to 7.82 in 1980 in the textile (spinning, weaving, knitting) subsector (see Table 7.6). Another consequence has been a decline in the ability to finance new investments in machinery, a step that may be needed for the survival of small and medium firms.

Table 7.6: AVERAGE DEBT EQUITY RATIOS IN TEXTILES AND CLOTHING, 1976-84 (%)

	Manufacturing	Textiles	Clothing
1976	364.6	492.2	505.7
1977	367.2	451.8	1,034.3
1978	366.8	496.8	1,117.8
1979	377.1	592.0	927.3
1980	487.9	782.0	1,586.6
1981	451.5	536.6	1,750.2
1982	385.8	530.2	2,155.2
1983	360.3	496.7	1,863.8
1984	342.7	480.6	

Source: Bank of Korea, Financial Statements Analysis.

Rising Wage Costs

7.24 The effect of rising wage costs varies across subsectors with respect to the share of labor costs in total costs and the extent to which real wages rise faster than productivity in each subsector. In the textile industry the most labor-intensive subsector is apparel. The production of yarns and fabrics from natural and artificial fibers are, on the other hand, relatively capital-intensive processes. It should also be kept in mind, however, that increases in wage costs in the cloth production process are fed through to the garment-making stage and affect factor proportions choices there also.

7.25 Real wages in the textile and clothing industries have generally risen at the same rate as average earnings in manufacturing over the last decade or so. On this count, therefore, no inter- or intraindustry trend is observable. However, both textiles and clothing experienced increases in wages over and above productivity increases through the 1970s (see Table 7.7) and this affected the industry's competitive position in the world. The average earnings per hour in spinning and weaving doubled during 1980-82 bringing Korea's wages up to par with its major East Asian competition and far beyond South Asian competitors (see Table 7.8). While some of these changes can be explained by changes in exchange rates[6/], the fact remains that rising wage costs have systematically eroded Korea's competitive position in all segments of the textile industry.

Table 7.7: NOMINAL WAGES AND PRODUCTIVITY GROWTH
(Annual % change)

	Nominal wages			Productivity	
	Textiles	Clothing	Manufacturing	Textiles	Clothing
1976	37	42	35	12	1
1977	24	28	34	4	15
1978	31	33	34	7	20
1979	32	32	29	12	17
1980	23	21	23	19	18
1981	19	20	20	21	33
1982	14	14	15	12	-18
1983	8	14	12	9	12
1984	7	10	8	8	3

Source: Wage data are from Economic Planning Board. Productivity data are from Korea Productivity Center.

6/ It should also be noted that in recent years contractionary macroeconomic policy has reduced the rate of nominal wage growth in Korean manufacturing below that of productivity growth while a steady depreciation of the won has also occurred. These policies have improved the competitiveness of Korean textiles and clothing, a development reflected in recent Korean trade performance.

Table 7.8: COMPARATIVE WAGE COSTS IN TEXTILES (SPINNING AND WEAVING)
(Gross earnings in US$/hr)

	1980	1981	1982
Netherlands	11.68	9.16	10.17
Fed. Rep. of Germany	10.65	8.17	8.38
US	6.37	7.03	7.53
Italy	9.12	7.23	7.06
France	8.57	6.40	6.36
Japan	4.35	4.90	5.64
UK	5.75	5.57	5.39
Korea	0.78	1.35	1.53
Hong Kong	1.91	1.42	1.40
India	0.60	0.69	0.66
Pakistan	0.34	0.42	0.37
Sri Lanka	-	0.16	0.32

Note: Changes in US dollar wage costs can also arise from exchange rate changes. Gross earnings include basic wage plus fringe benefits.

Source: Survey conducted by Werner International Management Consultants; cited in Cable, V. and Baker, B. (1983, Table 49).

7.26 Wages in textiles have been rising for both economy-wide and sector-specific reasons. Korea is now faced with a general shortage of unskilled labor, a shortage brought about by falling birth rates and rising education levels. In addition, workers are being attracted away from textiles, where working conditions are relatively harsh, into other sectors. Korea's high rate of economic growth has also created high rates of labor turnover and cross-sector mobility.[7/]

7/ The labor shortage and rising wage cost issue for textiles and clothing should not be confused with the recent concern regarding rising unemployment. The latter is a concern with a possibly temporary glut of university graduates arising from an increase in university enrollment rates four years ago. These graduates are not relevant for the blue-collar, low-skill, relatively low-pay jobs offered by the textile and clothing sector.

D. Private Sector Adjustment Strategy

7.27 The problems besetting the textile industry have not crept up suddenly but have been festering for several years. The textile industry has therefore had much time to examine the nature of its predicament and to undertake appropriate adjustments. The adjustment strategy that has emerged aims at changing the product mix, the cost and productivity relationships, the technology and the product market composition that characterizes the industry.

Product Mix Considerations

7.28 There is a consensus that the industry should move "upmarket" towards higher value-added, higher quality items. This should soften the consequences of increasing competition in the lower quality end of the market from lower-cost Asian textile producers as well as maximize the value of quotas placed on Korean exports in the US and the EEC. Moving up market will require investments in new machinery, in design facilities and in marketing arrangements. So far, progress in quality improvement appears to have been slow. The rate of replacement of old by new machines has been considerably below expectations (see Table 7.9). Some local textile companies have begun to develop their own patterns and fashion designs but the bulk remain dependent on buyer provided instructions or on standard or traditional designs. This places them at a disadvantage because demand in the upmarket segment changes frequently and buyers are wary of relying on producers who have little indigenous capacity to handle design changes. The industry is also hampered by the lack of marketing experience in this segment. Unlike in the market for standardized basic textiles, producers cannot always wait for buyers to come to them but must reach out and establish markets for themselves either through their own outlets or through close and quick connections with retailers.[8/]

Cost and Productivity Considerations

7.29 Since rapidly rising labor costs have been one of the industry's major problems there is an obvious need to control them. The industry has had limited success in controlling wage costs directly since the level is set by the economy at large. The high demand for labor outside textiles and the relatively poor working conditions within textiles limit the ability of the

---

8/ The product-mix choice is rendered difficult by several uncertainties. Fashions change quickly as far as apparel design is concerned. They may also change with respect to materials--witness the trend away from synthetics towards natural fiber in the 70s. Thus pursuing man-made fiber production in the future carries risks. On the other hand, improvements in technology have led to significant reductions in the cost of producing synthetic cloth and in the quality (feel, texture, ease of ironing, resistance to shrinking) of such cloth. Furthermore, the decline in the price of oil has also reduced industry costs. Investment choices with respect to product mix will have to contend with all the uncertainties of demand, input costs and production technology.

industry to cut labor costs directly. What has occurred instead is a retrenchment of labor from a level of 760,000 persons employed in 1981 to 720,000 persons in 1984. The bulk of the retrenchment has been in the spinning and weaving subsectors. Productivity has been a weak point in recent years. It has been plummetting in both clothing and textiles and in the former it has not been keeping pace with nominal wages. It would appear that the bulk of the adjustment in costs and productivity is yet to come, and that it will very likely mean reduced employment.

Technology Considerations

7.30 The desired changes in product mix and cost structure call for more reliance on small, flexible production technology as opposed to large, high volume machines. To some extent this puts such subsectors as spinning, weaving and knitting at a disadvantage since their present stock of machines is of the latter kind. At the same time technological developments in the garment industry have reduced the labor cost advantage enjoyed by Korea in the past. Adjustment in all subsectors will require an increase in the number of new, more advanced technology. Present indications are that little progress has been made on the technology front. Small producers have blamed the lack of credit availability for their failure to switch to the new machines. As far as larger firms are concerned, indications are that their profit margins have been so squeezed in recent years that they too have not had adequate resources to finance new machinery purchases. The rate of investment in the industry has been partly because of depressed profits, but also because of the great uncertainties that continue to characterize the future of the industry. Investment in new machines may not be enough if design and marketing arrangements are not also simultaneously improved. Not all firms will be up to the latter task and a period of shake-out and mergers may be necessary. This does not seem to have happened as yet as far as one can tell from data on the number of firms in the industry.[9/]

Product Market Diversification

7.31 Through most of the 1960s and early 1970s Korea textile exports went mainly to Japan and the US, these two countries absorbing as much as two-thirds of the total on average. The onset of more and more restrictive MFAs and bilateral quota agreements provided a spur to greater diversification through the 1970s as Korea sought new markets in the EEC, in the Middle East and in Hong Kong as well as in Africa and Latin America. By 1980 the individual shares of US and Japan had dropped considerably (to 22.2% and 17.6%, respectively) (see Table 7.11). Nevertheless, Korea's vulnerability to a reduction in demand from those two sources remains great. Further diversification will be constrained by three factors: (a) a considerable part of

---

9/ Data on comparative profitability (see Table 7.10) indicate that the textile sector (but not clothing) has consistently fared poorly relative to manufacturing in general since the mid-1970s. This suggests an additional reason for the lack of enthusiasm for capacity expansion in the textile business. The larger units in the business tend to be owned by conglomerates who may find it more profitable to divert their financial resources progressively into other industries.

the world remains in a state of low growth; (b) the Middle East market is no longer as lucrative as it has been in the 1970s; and (c) most of the OECD countries have raised high barriers to Korean exports and further growth in these markets is unlikely. Korea's problem is similar to that of other developing country textile exporters: the quota areas offer less scope for growth than before while the nonquota countries are generally too poor to allow much scope for growth.

Table 7.9: GROWTH IN EQUIPMENT INVESTMENT IN THE TEXTILE INDUSTRY, 1978-82
(% change in nominal values)

	1978	1979	1980	1981	1982	1983	1984
Textile industry	-16.6	49.8	-74.3	90.2	-4.3	28.3	19.0
Manufacturing sector	55.7	14.9	-24.4	-24.8	14.6	28.2	70.3

Source: Survey of Equipment Investment (December 1984), Korean Development Bank.

Table 7.10: PROFITABILITY TRENDS /a

	1975	1978	1980	1982	1983	1984	1973-77	1978-84
Textiles	0.25	3.69	-1.09	-0.66	0.59	1.19	3.2	0.6
Clothing	3.87	-0.55	0.99	2.79	5.08	5.46	3.8	1.8
Manufacturing	3.88	4.98	-0.23	1.03	3.27	3.41	5.3	2.3

Note: Profitability measured as ratio of normal profit to total assets.

Source: Financial Statement Analysis, Bank of Korea. Various issues.

Table 7.11: TEXTILE EXPORTS BY COUNTRY
(US$'000)

Country	1980	1981	1982	1983	1984
Total Textile Exports	5,014,323	6,185,807	5,924,541	6,050,879	7,078,571
U.S	1,113,480	1,468,201	1,619,003	1,981,146	2,585,108
Japan	884,987	1,018,528	974,102	721,660	1,018,927
EC	1,041,016	1,146,935	1,045,604	949,805	952,574

Source: Ministry of Trade and Industry.

E. Public Sector Role in Textiles

7.32 The Government's attitude towards the textile sector was very encouraging throughout the sixties and early seventies when it was the main foreign exchange earner and industrial employer. During this period the industry was supported by liberal credit facilities (through the nationalized banking system), generous rebates on import costs, and preferential tax and depreciation treatment. In the mid-seventies, however, the Government began to develop a heavy and chemical (HCI) industry sector in anticipation of a shift in Korea's comparative advantage, and to diversify the base for growth and exports. The implicit assumption was that Korea would lose competitiveness in light industries soon and that the light industry sector did not have enough steam left in it to power the country's future growth needs. The manner in which the switch to HCI was engineered proved detrimental to the textile sector. The two factors that most determine the industry's performance, capital and wage costs, were both sharply increased in the late seventies. Capital costs rose as the flow of relatively cheap bank credit to the sector dried up and firms were forced on to the curb market for larger and larger proportions of their financing needs. Wage costs rose as the massive investment in HCI and the inflation that was generated by this policy bid up the economy wide level of wages.

7.33 As of the early eighties, however, the Government has reduced its emphasis on HCI development and resurrected textiles as an important sector. Government policy towards textiles currently consists of the following: (i) contributions to a Textile Modernization Fund; (ii) capacity control regulations; (iii) preferential financial arrangements; and (iv) import protection and encouragement of R&D.

Textile Modernization Fund

7.34 This fund was set up in 1981 with the purpose mainly of providing loans with which to modernize factories. The main beneficiaries were intended to be the SMI's that form the bulk of the industry. The Government planned to contribute W 60 billion to the Fund over the period 1981-86 to match an expected contribution of W 60 billion from the industry itself and specific targets for the use of new machinery were set up. So far the performance of this Fund has not matched originial expectations. By mid-1985 only about W 27 billion had been collected and loaned--both Government and the industry have failed to provide the hoped-for contributions. The Government has been following a tight fiscal policy for macroeconomic stabilization reasons since 1981 and has not given the Fund a high enough priority to merit the release of resources from a tight budget. The industry has suffered from low profits in recent years and has apparently not pulled together in the way hoped for.

7.35 To some extent the Fund has suffered from poor design. Private industry contributions could not be expected to be large given that there is a heterogeneous mix of firms in the industry, of varying size and resources, with varying private adjustment strategies, and with varying prospects. The incentives to participate in the Fund probably differ significantly across these firms. At the same time the Government has not been in a position to bolster the Fund partly because of its fiscal policy and partly because of its professed intention, since 1981, to adopt a hands-off approach to industrial policy, especially at the industry-specific level. Furthermore, the Fund, even if fully capitalized, would be insufficient to meet modernization needs and hence private resources and initiatives must be relied upon ultimately. It is estimated that about W 600 billion is required to reduce the ratio of aged machines from present levels to a tolerable ratio of 20%.

Capacity Control Regulations

7.36 Government is now using an investment permit or licensing system to control investment in all segments of the textile industry except garments. It is not clear why the Government must control investment and capacity. If individual firms, in full knowledge of market conditions, still desire to expand capacity in a particular area and their banks are willing to finance them why should the Government stand in their way? The behavior of some firms may damage the profits of other firms but, to the extent that this is done by reducing costs and prices in the industry at large, there would seem to be a social benefit involved.[10/] Of course, some firms would have to close and some unemployment would result. But it is likely to be the relatively inefficient firms that go under. Also the resulting unemployment should be dealt with through retraining and redeployment programs and not by artificially propping up inefficient firms through guaranteeing their market share.

7.37 Perhaps capacity control could be exercised in a way that allows inefficient firms to exit the business gradually rather than suddenly. This has been the Japanese approach. In this case, a clear time limit should be specified for the duration of the capacity control regulation.

Textile Sector Financing

7.38 Currently the textile sector is being financed through regular commercial bank channels. Two special provisions affect the flow of finance: one applies to exporting firms and the other to SMIs. Since most textile firms are exporters and most clothing firms are SMIs these special incentives apply to the bulk of the industry. All exporters in Korea are given easier access to credit and can take advantage of easier financing terms (though not concessionary interest rates) and import tariff rebates. Similarly, SMI are given access to credit through a regulation that compels commercial banks to allocate 35% to 55% of their loans to SMIs. While the first set of incentives may translate into a significant source of advantage,

---

10/ This benefit is, of course, smaller to the extent that Korea's textile output is exported under quota arrangements such that domestic cost reductions simply translate into lower profits and lower foreign exchange earnings. Even in this case, however, a policy superior to capacity control would be to allow domestic competition to reduce costs and to slap on an export tax so as to capture the cost-reduction benefit domestically and prevent it from being passed on to quota-controlled markets abroad.

especially for larger firms who do most of the direct exporting, the latter set of incentives is not likely to be very effective. This is because the regulation covers all SMIs and not only textile and clothing units. As business prospects have deteriorated for textiles, banks have reduced their loans to this sector. Thus loans to the textile sector have declined from 31% of bank portfolio in 1974-76 to 23% in 1983-84. In particular, the Small and Medium Industry Bank has reduced its exposure to textile SMIs from 33% in 1976 to 23% in 1982 (see Table 7.4).

Import Protection for Textiles

7.39 Domestic producers are protected by high tariffs; these are likely to remain even as more and more items are removed from the list of import restricted items and therefore domestic producers will not feel the heat of competition for some time to come. The major competition is in low-grade or basic textiles and clothing and the main competitors are low-wage Asian producers such as China and India. The industry has been lobbying for continued protection in these market segments. At present the weighted average of ad valorem import duties on textiles and clothing is about 34% (see Table 7.12). The highest average duty is levied on clothing and made-up articles and fabrics (45% to 50%); lower duties are levied on fibers (15%)

Table 7.12: AVERAGE NOMINAL TARIFF LEVELS IN KOREA'S TEXTILE SECTOR (%)

	%
Clothing	50
Made-Up Articles	47
All	45
Industrial	50
Fibers	
All	15
Wool	27
Cotton	12
Man-made	30
Other	24
Yarns	
All	30
Wool	30
Cotton	30
Man-made	30
Other	30
Fabrics	
All	49
Wool	50
Cotton	50
Man-made	46
Other	50
Textiles and Clothing	34

Source: GATT (1984), Table 3.22.

and yarns (30%). Nontariff barriers are also present. These follow the pattern of intensity of tariffs in that they tend to keep out competitive labor intensive products such as garments.[11/] As of 1985, 105 out of 1,089 items connected to the textiles sector remained on the restricted import list. Government plans to reduce this number by 31, 32 and 19 in successive years from 1985 to 1987, by which time almost 98% of textile sector imports will be on a quota or restriction-free basis. However, some will still be subject to tariffs and special procedures.

R&D in Textiles

7.40 The Government promotes R&D work in textiles in three ways:

(a) support to the Korea Advanced Institute for Science and Technology (KAIST) budget for general R&D, including textiles;

(b) support/contributions to Korea Federation of Textile Industries (KOFOTI); and

(c) tax exemptions for R&D expenses to companies.

7.41 It is difficult to quantify the extent of government support for R&D work. It is clear, however, that Government attaches high priority to R&Ds work in general in Korea. It has been increasing its own expenditures on R&D since 1980. Also R&D by companies is on the rise even though it is still behind advanced countries. Thus R&D expenditure relative to total sales has risen from 0.33% in 1979 to 0.80% in 1982; however, this is below Japan's (at 1.65 in 1981).

F. Adjustment Strategies in Selected Countries

7.42 The textile industry in most OECD countries has been under pressure for over two decades now. In fact, during this time at least one country, Japan, has gone from having been a source of pressure for the others to an affected country itself. The problems in each case have been similar and have been connected essentially to the declining competitiveness of the domestic industry vis a vis textile exports from developing countries. The adjustment strategies followed have been one or more of the following:

(a) seek government support through protection (US, EEC) and subsidies (UK, Italy, France);

(b) seek to reduce production costs by modernizing equipment (UK, Fed. Rep. of Germany, Japan), by horizontal and vertical integration into

---

11/ Tariffs and purchase controls are also applied to man-made polyester fibers. This has caused some distress among local users. Apparently they have to buy domestically up to some specified level before being allowed to import.

mass-producing units (US, F.R. Germany, UK) and by use of immigrant labor and outward processing trade (F.R. Germany, France, US); and

(c) seek product innovation and process flexibility (Italy, Japan).

Governments have been deeply involved in the adjustment process through the extension of protection, subsidies, and special tax and depreciation measures.

UK Experience

7.43 The UK has had a long history of government assistance for the textile industry dating from 1959 when the Cotton Industry Act was introduced offering financial assistance for machinery modernization to increase productivity and to counter low wage imports from former colonies and from Japan. The rationale was that British textile firms, being mostly family run and not of large scale, would be unable to re-equip from their own (or the private sector's) resources and also that this would ward off demands for greater protection. This strategy was not successful. The uncertainty with regard to the competition from imports and the Government's attitude towards protection affected investment strategy with regard to retooling. The retooling that was done was not sufficient to restore competitiveness. A different approach was then emphasized. In the early 1960s, the private sector, led by Courtaulds and ICI, began restructuring through horizontal and vertical integration on a large scale but demanded higher "temporary" protection while the restructuring process was underway. The objective was to achieve economies of scale through integration and mass production of standardized items. Government bought this strategy and extended "infant industry" or, more appropriately, "born again" protection to the industry.

7.44 While a great deal of scrapping, re-equipping, merging and rationalising has been accomplished and while productivity has greatly increased, the intervention program must be judged a failure by the fact that the industry has not been rendered viable--there remains continuing pressure to retain and even to extend trade protection and other forms of assistance. Under cover of protection, the industry has become concentrated, it has laid off many workers and many companies have moved away from textiles into nontextile areas. It has not been rejuvenated through adjustment; it has been changed. The costs of financing this change have been very high. This type of adjustment, with its emphasis on process technology and lowering unit costs, may distract from more viable or efficient forms of adjustment such as the ones noted below for the Federal Republic of Germany (hereafter Germany) and Italy.

German Experience

7.45 The Federal Government has not generally intervened to help declining textile firms. Its major assistance to the sector has been in the form of acquiescence in EEC import regulations and MFA restrictions. More direct and substantial assistance has come from banks and regional governments. German banks often have equity positions (in addition to loans) in corporations and thus have more of an incentive to help firms survive. Regional governments are sensitive to local politics and often come to a sector's assistance if the employment and income impact is likely to be substantial.

7.46 Two sharply contrasting adjustment strategies have been followed in Germany. Most of the larger producers have attempted to adjust to competition by increasing capital intensity, sharpening product standardization and emphasizing volume of production. They have attempted to achieve economies of scale by vertical and horizontal integration and standardization and to reduce labor costs by using ultra-modern spinning and weaving technology. An alternative approach, followed by medium sized firms, has been in an almost opposite direction. Instead of integrating the firms have sought to become reorganized into smaller, flexible production units. Instead of focussing on volume these firms have sought to produce limited quantities of outputs with a heavy quality, design and marketing emphasis. However, these firms have also introduced a high degree of automation into their processing to avoid high labor costs.

7.47 Between these two strategies the former appears to have been less successful. The high volume/standard product strategy based on extensive automation has simply not been able to reduce labor costs enough to compete with Asian exports. The diffusion of new machines to competitors is fairly rapid and hence the advantage gained is short lived. Furthermore domestic wage costs have continued to rise. The alternative strategy has been more successful because it has exploited a comparative advantage based on the availability of good designers, the nearness to a consumer market in which quality is desired and fashions change quickly, and marketing ability in a market where needs must be created through advertising and satisfied through quick service.

7.48 Another element in Germany's adjustment approach has been the use of outward processing trade (OPT) whereby intermediate textile products have been sent abroad to lower wage countries on the Mediterranean rim (Yugoslavia, Tunisia) and in East Europe to be "made up" into apparel and semifinished goods and have been reimported into Germany for finishing and for final sales. The Government has helped by reducing tariffs on the import of goods under the OPT system.

7.49 The chief lesson of the German experience is that adjustment support, either government sponsored or private bank sponsored, to the industry segment that had lost comparative advantage was unsuccessful. West Germany has not been able to reestablish competitiveness in standardized basic textiles. It has been successful in precisely those market segments where, even without trade protection, further growth would have taken place. (The virtues of easy finance because of banking sector equity participation should not be overlooked).

The Italian Experience

7.50 Italy's experience reinforces the lessons drawn from the German experience. Italy has supported its textile sector in a far more interventionist and substantial fashion than Germany but the industry has been successful largely in those areas where a comparative advantage in design, quality and marketing could be exploited.

7.51 Government support for the industry has ranged from trade protection under EEC guidelines to specific subsidies to the industry such as (i) nationalization of large, ailing textile firms; (ii) concessional loans to firms through a Textile Modernization Fund; and (iii) rebates on labor costs arising from social security taxes and overtime payment rules. Nevertheless, the good performance of the Italian industry (it is the largest net exporter in the world) has been due largely to private initiatives. While the taken over firms remain a net drain on government resources to the tune of $100 million p.a. and while many large firms remain stuck with relatively inefficient labor forces kept in place by rebates on labor costs, small and medium firms have thrived. These firms have emphasized product quality and differentiation, flexibility of production and technical innovation in production. Smallness has been a source of advantage because it has allowed the flexibility in production that is required in making specialized, differentiated, high quality products the demand for which changes quickly. Smallness has allowed Italian firms to be run as family operations without labor unions and with low absenteeism. This has also allowed management decisions to be taken quickly and implemented swiftly to take advantage of shifting market fashions and demands.

7.52 The Italians have also been fortunate in having a textile industry concentrated in a few regions. This has allowed them to achieve central coordination (through cooperatives) in finance and marketing as well as in R&D. Nevertheless, the distinctive feature of the Italian experience has been private sector led decentralization and deconcentration in textiles and exploitation of market segments in which a comparative advantage could be seen. The Italian experience has been one of the success of small scale private enterprise amidst the failure of large scale nationalized units.

The US Experience

7.53 The US textile sector adjustment experience has been dominated by the prevalence, throughout the past two decades, of a high level of protection. From 1960 to 1980 less than 8% of domestic textile consumption was accounted for by imports. The industry has, therefore, responded not as much to competition from abroad but to internal changes in relative prices, to technological developments, and to domestic competition. Only very recently has import competition become the most important source of pressure and led to calls for even higher levels of protection. The clothing industry, however, has been under stress for a longer time despite high protection.

7.54 Over the years the textile industry has developed a high degree of product standardization and mass production. The existence of a large (and protected) domestic market has made it profitable to emphasize long runs and a small number of final products. Combined with a relatively modern stock of machines this sort of mass-production strategy has long kept American labor productivity higher than that of the European countries and Japan. Furthermore, pressure from rising real wages has been handled by a geographical shift of the industry from the unionized high wage areas of the northeast to the nonunionized lower wage areas of the south. The degree of unionization has also been lower than that found in the European countries. For all these reasons a mass-market strategy has been successfully followed in the US while it has failed to a large extent in the UK and in Germany.

7.55 While the dominant corporate strategy has been the one described above, the industry has not neglected product innovation and marketing. It has been a leader in the production of synthetic textiles and in establishing household textiles as a fashion sector. It has also featured substantial horizontal integration across the textiles and chemicals sectors. This combination, however, has not proved to be a successful one.

7.56 The evolution of the industry may also have been affected by exchange rate developments but there is no consensus on this point. It is thought that the weakness of the dollar relative to the European currencies throughout the 1970s kept the industry competitive (with Europe if not with East Asia). The subsequent rise in the value of the dollar in the 1980s may have had the opposite effect and led to an erosion of the industry's competitiveness. This may explain the industry's current protectionist fervor. The specific consequences of exchange rate changes for the industry as a whole are difficult to assess because of simultaneous changes in production technology and costs, variations in demand and competitors prices, and also because of process and product diversity within the industry.[12/]

7.57 Despite high average levels of labor productivity the US textile industry would probably not have fared well in direct, unprotected competition with East Asia. Protection has been important, perhaps even critical, to the industry's survival in its present form. In particular, the apparel sector and the mass-production yarn and fiber sectors have probably benefited enormously from protection. When this benefit is set against the cost that US consumers have had to pay to support the industry in this fashion it becomes obvious that protection has been extremely costly. According to one recent estimate, the efficiency loss to the US of restrictions on clothing imports amounted to $1.5 billion dollars (in 1980) or $170,000 per job saved. This compares with average labor compensation of $12,600 per year in this sector (Kalantzopoulos, 1986). Studies also indicate that society would have been much better off if adjustment assistance had been offered to the industry to promote a process of restructuring. Workers could have been retrained and re-employed in alternative occupations at far lower cost per head than that entailed by trade protection. Furthermore, while trade restrictions impose a continuing cost over the duration of the protection, adjustment assistance typically involves a one-shot grant or, at worst, obligations that decline over time as the adjustment progresses. Furthermore vigorous competition in the large domestic market available to American firms has offset some of the debilitating consequences of protectionism in reducing incentives to upgrade and innovate. Other countries, with far smaller domestic markets, cannot afford the luxury of protectionism in the long run.

12/ There are few rigorous studies of the effect of exchange rates on trade in specific products. One such recent study done by the USITC examined the impact of the dollars exchange rate on US trade in polyester staple fiber and denim fabric over the period 1977-82 and concluded that "although changes in exchange rates influenced trade, other trade factors were more important" (USITC, 1983, pg. 1).

The Japanese Experience

7.58 The Japanese began feeling the pinch of competition in the textile business towards the end of the 1960s first in their export markets (the US and the EEC) and very shortly thereafter in their domestic markets. Japan's share of world textile exports shrank from 10% to 5.5% between 1970 and 1980 as the East Asian trio (including Korea and Hong Kong) virtually ousted it from some segments of the market, particularly those characterized by labor-intensity (e.g., apparel) and/or standardized products (e.g., yarns and lower-grade fabrics). Identifying a secular rise in real wages as being the main cause of declining comparative advantage, the Japanese authorities undertook a series of measures designed to prod as well as assist the industry into adjusting in the direction of greater knowledge-intensity and higher productivity.

7.59 The prodding came in the form of a trade policy that gradually liberalized the import of textiles even in the face of industry opposition. Textile tariffs were substantially reduced through the 1970's and at present Japan's average textile tariff is lower than that of the US and the EEC. Quota restrictions have also been substantially reduced and Japan compares favorably in this regard also with the US and the EEC. However, while pressure was maintained on the domestic industry through trade liberalization adjustment assistance was also offered. Under the Special Textile Act (1967-74) this assistance took the form of (i) low interest loans to promote machinery modernization (scrapping and re-equipment), and (ii) special loans and grants to induce firms to merge (and reduce overall capacity) or to leave the industry altogether and set up in new lines of business. Under the New Structural Improvement Act (1979-84) the focus of attention shifted to small and medium textile industries and assistance was provided to help these industries switch to the higher value added segment of the market and to change lines of business. At the same time the emphasis on R&D was intensified for the industry as a whole. Finally, a major element of the Japanese adjustment strategy has been investment by Japanese companies in the textile sectors of their major competitors, primarily in East and Southeast Asia, so as to take advantage of the lower wages there and to survive as corporate entities if not as a national industry. Government has encouraged this trend.

7.60 The results of the Japanese adjustment experience may be summarized as follows. Firstly, the knowlege-intensive sector of the industry has continued to grow and prosper. Government support for R&D and the industry's move into R&D-intensive products appears to have paid off. Japan is now a major producer and exporter of certain types of synthetic fibers and fabrics. Japan's share of the US import market for textile (excluding apparel) has remained at around 20% for the last ten years largely because of a changing product mix based on a changing process mix. Second, the labor-intensive segment of the industry (primarily apparel) has shrunk domestically and virtually disappeared from overseas export markets. For example, Japan's share of the US import market for clothing shrank from 26% in 1963 to 11% in 1973 to only 3.1% in 1982. Its own imports of textile products (including apparel) jumped from $383 million in 1971 to $1,715 million in 1973 and proceeded to rise to over $3.8 billion by 1980. Its ratio of imports to domestic consumption has risen from less than 5% in 1970 to around 18% in the early 1980s. Third, the mass-producing yarn and fiber makers have either diversified out of the textile business or moved to overseas locations. Japanese firms are prominent in the production of synthetic fibers and yarns in particular in a number of East and Southeast Asian countries. While this move has not helped textile sector employment in Japan it has probably been the most productive use of Japanese textile capital. Fourthly, small and medium-sized firms have moved into the production of high quality high value-added items. Government help in marketing and financing has probably been of great use in retooling and reorienting this sector. In overall terms, the industry has shrunk in terms of output, employment and exports all of which peaked in the early 1970s. However, the reduction in size and relative importance has been managed without serious economic or social dislocation and there is certainly no sense of a permanent and continuing malaise concerning this sector in Japan as there is, for example, in several OECD countries.[13/]

G. Summary of Lessons of OECD Experience

7.61 The chief lessons of the adjustment experiences described above can be summarized as follows:

(a) While trade protection has lengthened the time available for adjustment and has maintained incomes in the textile sector, it has failed on two counts. It has not protected jobs in the industry and it has been extremely costly. Trade protection has not prevented the industry from shedding labor and from relocating away from the regions that were supposed to be the beneficiaries of the protectionist measures. By weakening the incentives to upgrade, innovate and invest, protection can become a problem rather than be a solution in the long run.

(b) Precisely those segments of the industry have prospered where the need for protection was least or, to put it another way, where a genuine comparative advantage was available. In Japan the knowledge-intensive segment of synthetic filament-based fibers and yarns has grown and prospered. In Italy and Germany the design and quality oriented segments have prospered while the mass-produced basic textile yarn segment, which has been the chief beneficiary of protection and subsidies, has not regained competitiveness.

(c) The high-volume, standard-product, mass-market strategy has been a failure despite heavy automation because of quick diffusion of

---

13/ The relative smoothness with which the Japanese industry appears to have adjusted should not be taken to mean that there have been no costs involved. Indeed the costs may have been high and may have fallen disproportionately on texpayers and consumers. While nominal textile sector tariffs and quota restrictions may appear to be modest the restrictive effect of other nontariff ("invisible") barriers may have been quite significant. The ratio of imports to consumption is far lower in Japan than in the EEC countries even if intra-EEC trade is excluded.

advanced technology to competitors and continuing rise in labor costs. The large scale vertically and horizontally integrated companies of the UK, France and Germany have been unsuccessful in achieving economic viability without government support. The large-scale units of the US have done relatively better but they would also not be competitive in a free trade environment.

(d) Small-scale units which have emphasized process flexibility, product diversity, and product quality have been very successful. The most lucrative segment of the product market has been the high quality fashion and design intensive segment. In order to be successful here decentralization and specialization of production units has been useful since this has allowed for flexibility to respond to fashion trend changes, small batch production to meet the limited (and changing) consumer demand for high quality, high fashion products, and for an emphasis on craftsmanship.

7.62 While it is obviously impossible to lay out a blueprint for success in adjusting to competition and changed circumstances, case studies allow a pattern of corporate strategies and behavior to be identified which is associated with a high degree of success. This pattern emphasizes flexibility and innovation. To quote from one review of such a pattern:

> "...successful textile or clothing firms in the developed countries usually exhibit some or all of the following features: professional management in an environment frequently characterized by family ownership; strong emphasis on cost and quality control, notably continuous modernization of the production process, internationalization of production and of marketing (OPT; foreign direct investment); a high standard of industrial engineering; substantial investment in training and motivating employees; emphasis on the high quality/high fashion/brand label edge of the market in both, design and marketing; and attractive "service package" for customers." (GATT, 1984, p. 171).

H. Future Prospects of Industry

7.63 The outlook for Korean textile and clothing exports is closely linked to the degree to which the developed countries will allow further inroads into their markets. In recent years they have shown little inclination to do so. A rather surprising degree of structural rigidity has characterized their economies through the past decade. This rigidity is partly explained by the fact that the textile and clothing industries are large employers and tend to be geographically concentrated and politically powerful. This means that their adjustment problems are likely to attract more attention than those of industries who may be hurt by measures taken to protect them. It is also the case that world demand for textiles and clothing has been growing at a slow pace, a fact related to the demographic dynamics of the developed countries, and hence it has been more difficult to make room or new producers. For example, while consumer expenditure on clothing increased at 4%, 4% and 7% for the US, the EEC and Japan, respectively, during 1963-73, the rates of growth for the next ten years (1973-82) were only 4%, 0.5% and 1%

respectively (Table 2.9, GATT). This is reflected also in the figures for growth of world trade in textiles and clothing. The average annual increase of textile and clothing exports declined from around 11.5% during 1963-73 to about 4.8% during 1973-82 (GATT, p. 36). While Korea has expanded its share of this trade in the latter, slower-growing period there is little hope that it can continue to do so in the future, especially considering the rise of protectionism and competition.

7.64 The history of the evolution of various restrictive arrangements concerning trade in textiles in clothing since 1962 does not lead to much hope for other than a narrow, domestic industry-oriented approach to the issue. Successive trade arrangements following the LTA and MFA I (i.e., MFA II and III) have been much more restrictive than the earlier ones. Moreover, in practice, these arrangements have been applied against exports from the developing countries rather than against mutual trade among the developed countries. What were intended to be temporary adjustment measures when first introduced have now been in force for 24 years and the clamor for further restrictions has not subsided. The scope for restrictions has been broadened by the introduction of the nebulous concept of "market disruption" a concept that has been applied in a fast and loose manner to violate, in spirit if not in letter, other agreements within the MFA protocol that are more favorable to exporting by developing countries [GATT, pp. 102, 103].

7.65 The extension of the MFA to 1986 contains provisions which enable importing countries to negotiate agreements which limit or eliminate greater growth and flexibility (i.e., carryovers, carry forwards and swings) from major suppliers such as Korea and Hong Kong. There are also provisions reaffirming less restrictive treatment for new suppliers (to the US) and small suppliers. This means that there will be little room for growth, especially in the major apparel categories, for Korea.

7.66 The US market, while large in absolute terms, offers little flexibility to the Koreans. The latest textile trade agreement, negotiated in 1982 and to run through 1988, covers 92% of all textile and apparel exports and provides only for a 2.5% p.a. growth subject to an average tariff of 22.3%. Under some protectionist bills now pending in Congress, US imports of Korean textiles would be cut by 35% (in some categories by 90%).

7.67 Korean textile exports may be able to keep growing by using third party quotas (i.e., by investing in Bangladesh and other countries) or by developing exports of non-MFA items such as leather gloves, silk fabrics, burlap fabrics among others. In 1980, $1.3 billion or 16% of US textile and apparel imports were not covered by MFA. There may be room for growth also in MFA-covered but less sensitive items where the share of imports to consumption is less than 30% and where, therefore, less strict monitoring applies.

7.68 There is also growing competition in Korea's main textile export lines from low-cost producers in Asia. The Chinese threat has already been documented. The crux of the matter lies in wage costs rather than in technology or in energy and capital costs. One can expect increasing mobility of textile technology and of capital in the future. Such mobility has worked to the disadvantage of the large-scale mass-production units of the UK and of

Germany in the past decade and Korea cannot expect to use technological sophistication alone as a source of future comparative advantage. Other countries can buy the same technology off the shelf also. Nor does Korea have any advantage in energy costs. While the recent decline in oil prices will certainly benefit Korea it will not necessarily increase its competitiveness relative to other textile and clothing exporters. Rising wage costs have already been referred to as a contributing factor to Korea's declining competitiveness in the late 1970s. While macroeconomic policies have helped rein in the explosive rate at which wages had been rising, the long term structural outlook is not favorable. Comparative population and labor force growth trends across Korea and her LDC competitors indicate that there will be more pressure on Korea's wages from the supply side than on those of her competitors. Consequently, Korea will continue to lose competitiveness in products with a low-skill and labor-intensive character.

7.69 While real wages have not been rising as rapidly in the eighties as they did in the seventies and while Korea has a tradition of good industrial relations in general, there is little room for maneuver on the wage front. There is little likelihood of competitiveness with other lower wage countries being enhanced by reducing real wages in Korea. If real wages begin to fall it is very likely that workers will leave this sector for others. The political consequences of falling real wages are also sobering. The most probable outcome is a reduction of employment in the textile sector together with an increase in real wages and in automation.

7.70 Pragmatic Korean adjustment to present and anticipated conditions in the global market for textiles would be to move in a direction consistent with the following goals/outcomes: (i) a change in the product mix of the industry in favor of higher quality and higher value added items, and (ii) a change in the factor proportions employed in the industry away from labor intensity towards greater mechanization and automation.

7.71 The desired change in product mix may require changes in production unit size. Since flexibility and variety of design are necessary for the production of high quality fashion items smaller spinning and weaving units are preferable to larger ones. At the same time such outputs will require sophisticated machinery for their production and hence high capital costs will have to be borne. Also, because such products rely on high pressure marketing for their success high per unit advertising and associated marketing costs will also have to be met. Finally, such products require expenditure on design which will also raise per unit costs. The ideal firm for such products would very likely be one which is large enough to meet the higher start-up and running costs and yet small enough to operate flexibly in response to constantly changing demand. Given access to credit some of the medium-sized firms in the business could switch to such new product lines successfully. Similarly, if the larger firms reorganized their businesses away from monolithic units producing standardized items and towards a number of smaller production units handling a variety of items, that too could work.

7.72 This path, emphasizing smaller firms, flexible production technology, and design and fashion orientation, has been successfully followed by Italy over the last two decades. Italy is the world's largest exporter of textile products and has thrived despite its wage-cost disadvantage vis-a-vis the East Asian countries by emphasizing its advantage in design and technology.

7.73 The upmarket move is not without its risks. Three may be most readily identified. First, there is limited room in the upmarket segment and manufacturers cannot move en masse into this area. Mergers, bankruptcies, retrenchment will necessarily be involved in the process. Second, there is much competition up there and it is of a nature that LDC's are not particularly strong in. Retailing networks are important. Advertising is important. Price is not the sole consideration. Hence production efficiency may be necessary but not sufficient. Third, the demand for upmarket products is highly income-elastic and, therefore, more prone to large swings over business cycles. Recessions hit upmarket firms hard (unless they are very, very upmarket).

I. Scope for Public Policy

7.74 The scope for industrial policy vis-a-vis the textile industry in Korea will be determined by (i) the Government's ability correctly to assess the nature of the industry's predicament and the directions in which the industry should move, and (ii) the costs and benefits of various kinds of government interventions to encourage changes in the necessary directions. Comparative experience from other countries suggests that governments are rarely well placed to assess and anticipate industry needs and long-run prospects. They react typically to representations from narrow industry groups who highlight problems and solutions in a selective and self-interested way. Typically the adjustment measures that they support turn out to be very costly and disproportionately beneficial to a few groups and, more often than not, fail in their attempt to block market forces. This is probably the chief general lesson of the experience with textiles of the EEC and US governments.

7.75 Guidelines for appropriate industrial policy vis-a-vis the textile sector in Korea may be drawn from the experience of the OECD countries. However, they must be modified to take into account the fact that Korea's predicament is partly due to a decline in competitiveness and partly to an increase in protectionism. Korea's policy towards textiles is currently a blend of support for process innovation, trade protection, preferential financing and capacity control.

7.76 The support for process innovation takes two forms, access to credit given through the Textile Modernization Fund and indirect support through encouragement of R&D work. It may be recalled that direct government support of textile process innovation in the OECD countries has been unsuccessful in restoring industry viability or maintaining jobs and market shares. Part of the problem there has been support for a high-volume/standard-product strategy which was undermined by continuing wage increases, and the quick adoption by competitors of the advanced technology. Korea may be more successful with this strategy if labor cost increases can be kept in check and to the extent that its major competitors are not yet in a position to use advanced technology. Since the strategy's outcome is uncertain there should not be any undue encouragement of it. The speed at which to scrap and re-equip and the type of new equipment chosen should be left largely to private initiative. The Government's policy of modest contributions to the Textile Modernization Fund is, under the circumstances, a prudent one. Industry has taken a wait-and-see approach to retooling. Government should not rush in where industry has feared to tread.

7.77 Tax incentives for firm-level R&D work are both desirable and helpful. Many of the uncertainties in the field can be reduced through R&D work and since the social benefits are usually greater than the private benefits government support for this sort of work can be justified on externality grounds. Promotion of R&D should help in matters of machinery use and adaptation as well as in product design and quality enhancement. There are likely to be positive externalities also in government support of training facilities for textile design and for marketing. The industry is likely to benefit more than any single firm from improvements in the quality and image of Korea's textiles.

7.78 The OECD countries have chosen to protect their textile sectors from competitive imports by raising tariff and quota barriers. This has been an extremely costly move. Job-saving has been minimal since firms have used the rents from protection to install labor-saving machinery or, in some cases, to move into non-textile products. All in all trade protection has proved to be always costly and often ineffective. Furthermore protection requires a large domestic market if its cost is to be reduced. Korean policy towards the import of textiles from lower cost countries should be framed with this lesson in mind. Should greater import protection become necessary for political reasons the best procedure would be to implement a declining tariff. Of course, it is hard to argue for free trade when export markets are being closed by protection. Nevertheless, for small economies the best solution is usually to remain open even if some of their partners are moving away from free trade. Korea can ill afford a textile sector that is a pensioner of the Government. The best way to deal with the loss of jobs that would occur in the sector were import to be considerably liberalized would be to ease the movement of displaced workers into new occupations and industries. A sound macroeconomic policy combined with some government-sponsored training programs should prove to be the socially least-cost way of managing the textile employment problem.

7.79 Credit availability has been a factor in the ability of small and medium firms to prosper in the textile sector of Italy and Germany. It is likely to be critical in the future development of the Korean industry also. In this regard the Government's policy of reserving a portion of bank credit for SMIs is a step in the right direction. At this stage in the development of Korea's financial markets a certain amount of positive discrimination in favor of SMIs is desirable. Moreover, the measure is not industry-specific and hence not as likely to misallocate resources. However, the financing of the industry must be subject to rational financial decision-making.

7.80 It must be kept in mind that the optimal-size question remains unresolved. While the EEC experience has revealed the strengths of small to medium firms it is possible that future technological developments in textiles may prove to be more supportive of large scale units, especially if synthetic material based production continues to grow or if the sewing process is further automated. Furthermore, even in the EEC experience, centralized financing and marketing (through cooperatives and regional banks) have played an important role. Given the uncertainties regarding optimal size the Government should not influence the scale decision unduly through credit allocation policies or through capacity control legislation. The Government's current scale-based policies should be re-examined at regular intervals in order to prevent serious distortions from accumulating.

7.81 The goal of public policy should not be to preserve capital and employment in the industry at all costs but to ease adjustment towards other uses of the industry's factors of production. This could change the character of the industry (in terms of product mix and factor proportions, for example) but that is, in itself, not an undesirable consequence. The important thing is to undertake measures which reduce the overall costs of adjustment. This requires that the interests of sectors other than textiles and clothing also be kept in mind and the notion of a shifting comparative advantage be accepted and indeed converted to one's benefit. Public policy should not be used to block the signals being given by the international marketplace but rather to make use of them.

Table A7.1: THE RELATIVE IMPORTANCE OF THE TEXTILE INDUSTRY

Function	Category	Unit	1971	1976	1982	1983	1984
Value-added	GNP (A)	W bln (1980 prices)	18,797	29,804	41,737	45,718	49,180
	Manufacturing sector (B)	"	2,918	6,974	11,933	13,235	15,172
	Textile industry (C)	"	458	1,304	1,899	2,007	2,089
	C/A	%	2.4	4.4	4.5	4.5	4.2
	C/B	%	15.7	18.7	15.9	15.2	13.8
Exports	Total (A)	US$ mln	1,068	7,715	21,616	24,445	29,245
	Textile products exports (B)	"	452	2,740	5,925	6,051	7,079
	B/A	%	42.3	35.5	27.4	24.8	24.2
Employment	Total economy (A)	Thousand persons	10,066	12,556	14,424	14,515	14,417
	Manufacturing (B)	"	1,336	2,678	3,047	3,275	3,351
	Textile industry (C)	"	163	641	746	736	738
	C/A	%	1.6	5.1	5.2	5.1	5.0
	C/B	%	12.2	24.0	24.5	22.5	22.0

Source: Bank of Korea and Korea Federation of Textile Industries (KOFOTI).

Table A7.2: PRODUCTION OF TEXTILE YARNS, FABRICS AND CHEMICAL FIBERS

Classification	Year	1979	1980	1981	1982	1983	1984
Yarns (M/T)	Cotton yarn	349,610	381,386	383,372	403,608	420,681	437,762
	Worsted yarn	21,181	17,940	23,964	23,116	51,833	27,970
	Woolen yarn	15,290	17,487	17,024	18,769	22,563	22,866
Fabrics ($1,000\ m^3$)	Cotton fabrics	940,386	1,016,528	1,032,502	982,813	1,016,046	1,315,139
	Worsted fabrics	34,428	36,020	33,660	35,457	36,190	40,175
	Woolen fabric	15,222	13,525	11,577	10,190	11,450	12,634
	Silk fabrics	49,828	58,066	52,044	49,911	44,327	41,007
	Rayon fabrics	109,466	127,566	62,624	34,414	17,341	18,371
	Synthetic fabrics	680,470	792,566	914,822	1,028,690	1,133,396	959,046
Chemical fibers (M/T)	Nylon	114,599	119,897	124,260	119,490	123,928	133,283
	Polyester	230,479	277,080	332,073	338,713	387,586	451,094
	Acrylic	132,159	139,436	153,847	154,867	152,304	162,660
	Viscose	21,541	25,438	24,673	13,594	10,749	10,569
	Acetate	7,758	7,466	7,910	8,067	8,164	8,088
	Others	7,685	3,892	3,123	2,100	554	952
	Total	514,321	573,214	645,886	636,831	683,285	766,646

Note: Cotton Yarn and Fabrics include Cotton Blended Yarn and Fabrics
Worsted Yarn and Fabrics include Wood Blended Yarn and Fabrics.

Source: KOFOTI.

Table A7.3: SPINNING FACILITIES AND CAPACITY FOR CHEMICAL FIBERS

Classification	Year	1979	1980	1981	1982	1983	1984
Spinning facilities (spindles)	Cotton spinning	3,074,892	3,167,124	3,192,212	3,128,809	3,243,040	3,251,342
	Worsted spinning	713,044	823,924	839,080	837,216	895,412	849,036
	Woolen spinning	98,630	111,996	115,264	123,890	131,059	144,837
	Total	3,886,566	4,103,144	4,146,556	4,089,915	4,269,511	4,245,215
Capacity for chemical fibers (MT/D)	Nylon	262.5	262.5	264.3	374.0	384.0	394.8
	Polyester	495.0	580.5	611.0	1,0[illegible]1.6	1,054.1	1,132.6
	Acrylic	320.5	345.5	345.5	439.0	459.0	419.0
	Viscose	59.2	59.2	109.2	109.2	109.2	109.2
	Acetate	15.5	15.5	15.5	22.5	22.5	22.5
	Others	26.0	29.9	23.5	21.8	9.5	9.5
	Total	1,178.7	1,299.7	1,369.0	2,008.1	2,038.3	2,127.6

Source: KOFOTI.

Table A7.4: TEXTILE FACILITIES FOR LOOMS, KNITTING MACHINES AND OTHERS
(Unit: Set)

Classification Year	1979	1980	1981	1982	1983	1984
Cotton loom	52,768	55,179	51,699	47,857	45,448	49,489
Filament loom	84,337	85,286	85,709	84,976	74,387	65,118
Wool loom	3,759	3,439	3,555	3,084	3,890	3,154
Embroidery loom	448	448	294	318	300	204
Towel loom	1,808	1,935	1,882	2,086	2,117	2,156
Knitting M/C	52,735	43,085	48,311	41,129	41,445	41,486
(Circular knitting)	18,172	14,871	20,378	13,683	12,017	12,045
(Flat knitting)	22,265	15,026	14,047	15,639	15,202	15,705
(Warp knitting)	560	1,008	2,605	810	808	463
(Socks & stockings)	7,899	8,364	7,946	7,680	9,344	9,520
(Gloves)	3,849	3,816	3,335	3,316	4,074	3,753
Sewing M/C	225,642	275,414	321,694	359,308	359,781	318,736
Dyeing M/C	3,778	3,621	4,122	4,043	3,975	4,025
Twisting M/C (spindle)	155,000	158,500	167,500	155,804	134,000	125,000

Source: KOFOTI.

Table A7.5: DEMAND AND SUPPLY STATUS
(Unit: M/T)

Classification		Year	1979	1980	1981	1982	1983	1984
Input	Import	Fibers	370,931	390,210	391,165	411,659	435,121	451,096
		Yarns	36,682	25,335	41,137	46,510	43,960	45,592
		Fabrics	33,803	33,099	33,889	31,700	42,029	44,684
		Clothing	1,036	744	504	269	112	220
		Others	100	94	59	54	94	181
		Subtotal	442,551	449,482	446,754	490,192	521,316	541,773
	Production	Man-made fibers	562,574	617,613	696,056	680,571	741,420	830,362
		Natural fibers	11,870	12,990	9,500	6,945	5,372	5,062
		Subtotal	574,444	630,603	705,556	687,516	746,792	935,424
	Total		1,016,995	1,080,085	1,172,310	1,177,708	1,268,108	1,377,197
Output	Export	Fibers	12,457	34,453	34,408	26,735	26,047	73,631
		Yarns	123,472	178,605	170,232	153,731	184,261	193,276
		Clothing	188,816	178,955	230,068	230,425	251,028	242,877
		Others	1,114	1,802	2,588	2,923	2,067	2,963
		Subtotal	543,278	666,518	754,051	706,702	798,476	848,279
	Domestic consumption		339,325	340,062	335,106	358,419	370,771	388,280
	Inventory & others		134,403	73,505	83,153	112,587	98,861	140,638

Source: KOFOTI.

Table A7.6: TEXTILE EMPLOYMENT BY SECTOR
(Unit: Persons)

Sector	Year	1980	1981	1982	1983	1984	1984/ 1983 (%)
Total		731,963	760,333	746,012	735,620	719,357	97.8
Garment		368,013	382,156	383,355	384,448	383,936	99.9
Spinning		118,905	117,063	111,584	112,374	1091294	97.3
Weaving		97,840	99,561	94,901	91,543	80,664	88.1
Knitting		60,030	54,839	51,910	51,551	59,520	115.5
Others		87,175	106,714	104,263	95,704	85,943	111.4

Source: KOFOTI.

Table A7.7: SIZE STRUCTURE OF THE TEXTILE INDUSTRY, AS OF END OF 1983

	Textiles		Clothing		Textile industry	
	Small & medium	Large	Small & medium	Large	Small & medium	Large
No. of establishments	5,454 (97.4)	144 (2.6)	3,343 (96.4)	125 (3.6)	8,797 (97.0)	268 (3.0)
No. of employees	215,403 (56.3)	167,056 (43.7)	135,002 (60.5)	88,313 (39.5)	350,405 (57.8)	255,369 (42.2)
Output (W bln)	2,373.7 (39.0)	3,713.1 (61.0)	951.2 (43.1)	1,256.5 (56.9)	3,324.9 (40.1)	4,969.6 (59.9)
Value added (W bln)	1,001.9 (44.8)	12,032.5 (55.2)	456.1 (50.9)	439.4 (49.1)	1,458.0 (46.6)	1,671.9 (53.4)

Note: 1. The survey covered all business establishments with at least 5 workers. Figures in parentheses denote % relative shares.
2. Small and medium industries are those industries that employ less than 300 employees.

Source: Report on Mining and Manufacturing Survey, 1985.

Table A7.8: GROWTH IN TEXTILE EXPORTS, 1976-84
(US$ mln)

	Textiles		Clothing		Total textile /a	
1976	954.4		1,845.5		2,740.1	(46.5)
1977	1,081.7	(13.3)	2,061.7	(11.7)	3,039.3	(10.9)
1978	1,533.3	(41.7)	2,574.7	(24.9)	3,981.9	(31.0)
1979	1,814.5	(18.3)	2,849.4	(10.7)	4,501.3	(13.0)
1980	2,197.5	(21.2)	2,946.9	(3.4)	5,014.3	(11.4)
1981	2,449.7	(11.5)	3,862.7	(31.1)	6,185.8	(23.4)
1982	2,237.7	(-8.7)	3,773.9	(-2.3)	5,924.5	(-4.2)
1983	2,413.4	(7.9)	3,707.3	(-1.8)	6,050.9	(2.1)
1984	2,601.5	(7.8)	4,499.5	(21.4)	7,078.6	(17.0)

Notes: Figures in paretheses denote annual growth rates.

Source: Bank of Korea, Economic Statistics Yearbook; and Ministry of Commerce and Industry.

Table A7.9: **AGE OF TEXTILE MACHINES (1980)**

	Over 20 years	Over 15 years	Over 10 years	Under 10 years	Total
Spinning machines	497,454	720,941	1,326,175	2,936,009	4,262,184
	(11.7)	(16.9)	(31.1)	(68.9)	
False twisters	838	2,154	70,416	143,932	214,348
	(0.4)	(1.0)	(32.9)	(67.1)	
Dyeing machines	363	1,100	3,275	3,571	6,846
	(5.3)	(16.1)	(47.8)	(52.2)	
Looms	17,293	27,354	59,402	130,233	189,635
	(9.1)	(14.4)	(31.3)	(68.7)	
Knit machines	2,107	9,217	38,477	35,366	73,843
	(2.9)	(1.5)	(52.1)	(47.9)	
Sewing machines	2,892	14,662	54,911	109,638	164,549
	(1.8)	(8.9)	(33.4)	(66.6)	
Embroidery machines	20	221	1,313	945	2,258
	(0.9)	(0.8)	(58.1)	(41.9)	

Note: Figures in parentheses denote % ratio to the total.

Source: KOFOTI.

Table A7.10: AGE OF TEXTILE MACHINES IN FEBRUARY 1983

	Over 20 years	Over 15 years	Over 10 years	Under 10 years	Total
Spinning machines	661,730	1,064,031	1,881,019	2,355,739	4,236,758
	(15.6)	(25.1)	(44.4)	(55.6)	
False twisters	4,224	34,346	135,264	71,462	206,726
	(2.0)	(16.6)	(65.4)	(34.6)	
Dyeing machines	894	2,366	4,237	2,843	7,080
	(11.2)	(33.4)	(59.8)	(40.2)	
Looms	19,604	40,246	89,463	86,332	175,795
	(9.9)	(22.9)	(50.9)	(49.1)	
Knit machines	6,306	27,962	47,577	16,120	63,697
	(9.9)	(43.9)	(74.7)	(25.3)	
Sewing machines	15,080	37,599	88,110	69,7908	157,900
	(9.6)	(23.8)	(55.8)	(44.2)	
Embroidery machines	71	284	1,102	288	1,390
	(5.1)	(20.4)	(79.3)	(20.7)	

Note: Figures in parentheses denote % ratio to the total.

Source: KOFOTI.

CHAPTER 8: ELECTRONICS IN KOREA: A CASE STUDY OF AN EMERGING INDUSTRY

A. Introduction

Overview

8.01 The electronics industry in Korea dates back to 1959 when firms began assembling radios from imported parts and components. Since then it has grown more rapidly than any other industry in the country and has become highly diversified, its products classifiable into three subsectors: (i) consumer electronics; (ii) industrial electronics; (iii) electronic parts and components. Traditionally, Korea's electronics industry has been heavily oriented toward consumer electronics, in particular, television receivers and audio equipment as well as semiconductor assembly for export. The industrial electronics subsector (computers and peripherals, communications equipment, instruments and controls) has represented a small share of total industry output, however, in recent years that share has been increasing markedly, as a number of firms, building upon their experience in consumer electronics, have been diversifying into industrial electronics production. While communications equipment constitutes the bulk of such production, production of computers and peripheral equipment has also been expanding dramatically. Although Korean computer production remains largely oriented toward low-end peripherals like dumb terminals and monitors, there has been a trend recently toward the production of more sophisticated products, including whole systems. For example, a number of Korean electronics firm have been successfully marketing IBM PC-compatible systems in the United States, either under their own brand names or as an OEM (original equipment manufacturer) supplier to the US firms. Within the broad category of telecommunications equipment, there has been an effort in recent years to upgrade technologically, for example, by designing and developing an indigenous digital publc switching system as well as by acquiring the know-how to manufacture fiber optic-based systems and components.

8.02 There has been a parallel trend in the component subsector toward the production of more sophisticated components as well as the introduction of more complex processes. The most outstanding instance of this has been the large scale investment by Korean firms in semiconductor wafer processing since 1983. From 1974 through 1981 cumulative fixed investment in semiconductor operations amounted to US$211 million. In 1983 alone it totalled US$300 million, then climbed to US$469 million in 1984 and was expected to reach US$487 million in 1985. (In 1986 it is expected to decline slightly to US$412 million.) Unlike in the past, the overwhelming share of new semiconductor investment has gone into water fabrication as opposed to assembly. This decision to place a high priority on the acquisition of semiconductor technology has exposed the Korean industry to probably the most serious competitive risk of its history. Given the potential importance of semiconductor technology for the future development of Korea's electronics industry, it is valuable to examine the Korean industry's strategy in this area in more detail (see Section B below).

8.03 The explicit thrust of Korean government policy toward the electronics industry since at least 1982 has been to promote the development of a technologically more advanced and autonomous industry. While the GOK recognizes the critical importance of continued access to foreign technology for the Korean industry's development, it also realizes that indigenous technology development not only enhances Korea's ability to absorb foreign technologies but also makes foreign firms more willing to transfer their state-of-the-art technologies to Korea. To what extent the technological upgrading of the electronics industry would occur without the implicit promotion of the government is an important question, although the recent foray of Korean semiconductor firms into very large scale integration (VLSI) memory production has occurred for the most part independently of government initiative.[1/] In the Korean industry today, the Government is confronted with a situation of having to decide, ex post facto, whether to lend its support to private sector initiatives, to adopt a laissez-faire approach, or actively to discourage firms from pursuing the high risk strategy some of them have chosen.

8.04 Despite Government's reorientation in favor of functional industrial incentives and greater neutrality, its continuing role in the domestic financial sector as well as its policies towards conglomerates will put it to some extent into the future path of the electronics industry. For that reason, it is valuable to investigate the characteristics of that industry in further detail.

Characteristics of the Industry

8.05 The output of the electronics industry in 1984 totalled US$7.17 billion, 58.6% of which was directly exported and another 10.9% of which was incorporated in end products destined for export. Between 1970 and 1984 the industry grew at an average annual rate of 35%. The output structure of the industry has undergone a marked change over the period. In 1970, electronic components accounted for 55.7% of output and consumer electronics 28.3%. Industrial electronics made up the remaining 16%. By 1979, the share of consumer electronics had risen to 41.9% and that of components had fallen to 48.4% (see Figure 8.1). The industrial electronics share, meanwhile, had slipped to 9.7%. After 1979, a significant reversal began, with the consumer electronics share shrinking and that of industrial electronics rising. By 1984, consumer electronics accounted for only 33.8% of output while industrial electronics had slightly surpassed its 1970 share. For purposes of comparison, Table 8.1 contains a breakdown of the output structures of the electronics industries of Japan, US and Europe. Japan has an intraindustry structure most like Korea's, although the marked difference is the much higher share of industrial electronics in total output in Japan. Conversely, Korea has a much higher component share than any of the other producers, reflecting the heavy weight of semiconductor assembly in Korea's overall output structure.

1/ The Government did have some role to play in early experiments with commercial scale wafer fabrication, and it still plays a small role in supporting joint research efforts with private industry.

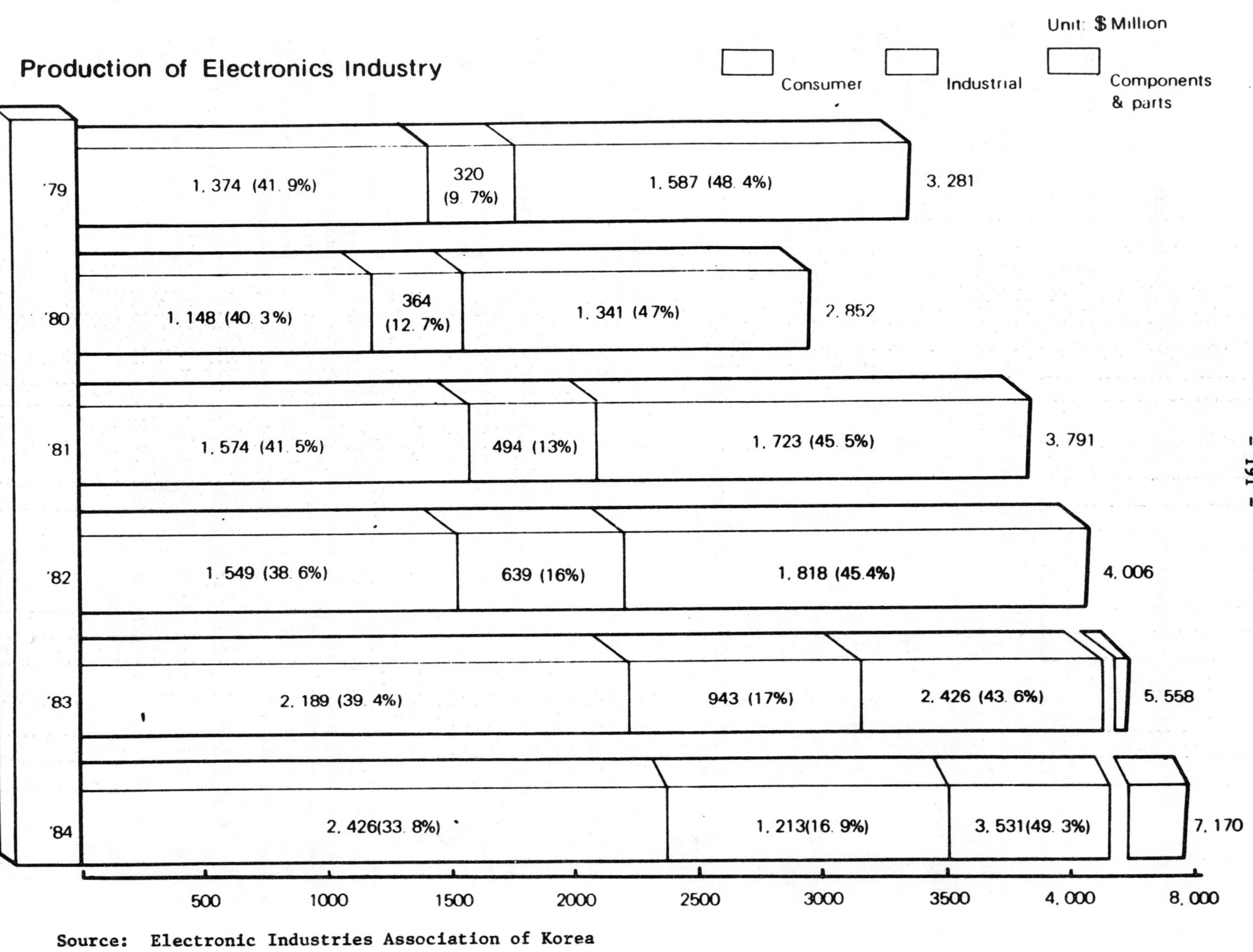

Source: Electronic Industries Association of Korea

Figure 8.1

Table 8.1: OUTPUT STRUCTURE OF THE ELECTRONICS INDUSTRY

	Korea		Japan		US		Europe	
	1978	1984	1978	1983	1978	1983	1978	1983
TOTAL (US$ m.)	2,271	7,170	30,063	53,405	64,944	143,831	41,587	57,782
of which:(%)								
Consumer	40.8	33.7	34.7	30.2	14.3	9.9	27.4	23.8
Industrial	9.2	16.9	36.3	36.3	68.1	68.6	52.8	58.3
Components + parts	49.9	49.2	29.0	33.5	17.5	21.5	19.7	17.9

Source: Electronics Industry Today and Tomorrow, 1985 (Electronic Industries Association of Korea).

8.06 Nevertheless, within the component subsector a noticeable restructuring has occurred over the last decade and a half. In 1971, semiconductor devices (transistors and integrated circuits) accounted for 56% of total component production and two thirds of exports; by 1979, the semiconductor production share had fallen to 29% and the export share to one half (see Table 8.2). By 1984, the semiconductor production share had risen again, but to only 36% of total component production; semiconductor exports had risen somewhat more steeply, to 61% of component exports. In large part the performance of semiconductors in 1984 reflected buoyant world market conditions. In any event, the slight recovery in the production and export shares of semiconductors does not negate the dominant trend in the component sector toward higher levels of integration with domestic equipment manufacture. In particular, the rapid growth in consumer electronics production during the 1970s stimulated local demand for electronic tubes, capacitors, resistors, transformers, audio parts, TV tuners, etc. As the Korean industry moves toward integrated semiconductor production, and if Korean firms should become major exporters of memory chips, the IC share of component production and export may well continue to rise through the remainder of the decade.

The Role of Electronics in the Economy

8.07 Korea's electronics industry contributes significantly to overall economic--and in particular to industrial--activity. Table 8.2 contains a summary of certain key economic ratios for Korea's industry, as well as the comparable figures for Japan.

Table 8.2: KEY ECONOMIC RATIOS OF KOREA'S ELECTRONIC INDUSTRY, 1983

	Korea	Japan (1982)
(1) Electronics industry output as % of mfg. output (W 4,036,134 mln)	6.67	4.55*
(2) Electronics industry value added as % of mfg. VA	7.24	8.78
(3) Electronics industry VA as % of GDP (W 1,514,093 mln)	2.54	2.48
(4) Electronics industry employment as % of mfg. employment (187,628)	8.33	8.91
(5) Electronics exports as % of mfg. exports	13.07	15.69*
(6) Electronics exports as % of industry output	58.56	51.25

* 1981 data.

Source: Report on Mining and Manufacturing Survey

The most striking difference between the Korean industry and that of Japan is the share of value added (VA) in output. From Table 8.1 it is apparent that the VA/output ratio in Korea is significantly lower than that in Japan. This is explainable in large part by the different degrees of import dependence, i.e., whereas Korea electronics imports amount to roughly 46% of production, in Japan it amounts to only around 6%. The high import dependence of the Korean industry is attributable in considerable measure to the sizeable semiconductor assembly industry which still imports the bulk of material input requirements. In addition, electronic equipment manufacturers still import most of their semiconductor needs, though that may change with the expansion of domestic wafer processing activities. By one estimate,[2] Korean firms must still import 80% or more of their microelectronics and other sophisticated parts requirements. A second explanation for lower VA/output ratios in Korea may be the lower R&D intensity of output vis-a-vis Japan. By producing products at an earlier stage of the product cycle, Japan may be able to earn larger rents.

2/ US Department of State (Commercial Section, US Embassy, Seoul).

Table 8.3: KOREA: ELECTRONICS EXPORTS, IMPORTS AND TRADE BALANCES BY SUBSECTOR, 1984

	Exports	Imports	Trade Balance
Consumer Electronics	1,523	174	1,349
Industrial Electronics	552	892	(340)
Electronic Components	2,129	2,097	32
Total Electronics	4,204	3,163	1,041

Sources: Statistics of Electronic and Electrical Industries; Import Statistics of Electronic Products and Electrical Appliances (Electronic Industries Association of Korea, 1985).

8.08 Despite the import dependence of Korea's electronics industry, the industry is a net exporter. In 1984, for example, Korea registered net electronics exports of $1.04 billion. Consumer electronics accounted for the bulk of the trade surplus. Table 8.3 summarizes trade data for 1984 by subsector. The industrial electronics subsector continues to be a major net importer, though the size of the deficit in relation to imports has been declining over time. Whereas in 1976 the deficit on industrial electronics trade came to 45% of imports, by 1984 the ratio had fallen to 38%.

8.09 While exports have been a critical source of demand for the output of Korea's electronics industry, by no means have they been the only one. The domestic market for electronic equipment has grown at the average rate of 32% a year between 1972-84 (see Table 8.4), compared to a 39% per year growth in exports. At the same time, the domestic market share of Korean-made electronic equipment has been rising considerably. Whereas in 1972 imports still accounted for 58% of equipment sales in Korea, by 1984 the import share had declined to 41%. The reduced import dependency was most dramatic in the case of consumer electronics, where the local production share increased from 45% to 84% between 1972-84.

Table 8.4: DOMESTIC AND EXPORT MARKET SALES OF ELECTRONIC EQUIPMENT (US$ m)

	1972		1984		CAGR(%)	
	Domestic	Export	Domestic	Export	Domestic	Export
Consumer	44	35	1,077	1,523	30.5	36.9
Industrial	53	4	1,553	552	32.5	50.8
Total	97	39	2,630	2,075	21.7	39.3

Source: Based on production and trade statistics of Electronic Industries Association of Korea.

8.10 In terms of employment, the absolute contribution of the electronics industry to manufacturing sector employment is quite significant. Due to a combination of domestic and world market conditions, electronics employment began falling after 1978. By 1982 some 30,000 jobs had been eliminated, a 16.3% decline from the 1978 level of 183,635 employees. (Production increased over the same period by 76%.) Employment rose steeply again in 1983 and 1984, reaching a level of 218,000 workers by the latter year. By 1985 employment was approaching 300,000. The structure of the workforce has undergone a dramatic shift since 1978. Whereas in that year factory workers accounted for nearly 80% of total employees, by 1982 this share had fallen to 55%. Meanwhile, the share of engineers and technicians in total employment rose from 18% to 27%. Much of that shift is attributable to rising automation levels, especially in semiconductor assembly. The increasing R&D intensity of Korean electronics has also contributed to the shift toward skilled labor. The trend toward a higher ratio of engineers and technicians to direct production workers is expected to continue through the early 1990s at least (see Table 8.5).

Table 8.5: EMPLOYMENT BY ELECTRONICS INDUSTRY, ACTUAL AND PROJECTED
(1,000 persons; %)

	1981	1986 (projected)	1991 (projected)
Specialized Engineers (including R&D staffers)	12 (4.7)	62 (10.9)	114 (15)
Engineers	54 (21.8)	113 (19.9)	190 (25)
Skilled Workers	143 (57.9)	300 (52.9)	357 (47)
Others	38 (15.6)	92 (16.3)	99 (13)
Total Number of Employees	247 (100)	567 (100)	760 (100)

Source: Electronic Industries Association of Korea

Global Developments

8.11 The redirection of Korea's electronics industry has been precipitated by a combination of changing world market conditions, shifting perceptions of dynamic competitive advantage and explicit government policy. Since Korea's electronics industry has historically been and continues to be export-oriented, global developments have an especially significant impact on the structure of the domestic industry. One important factor accounting for the relative deemphasis of certain consumer electronics products has been the imposition of restrictions on imports of Korean consumer electronics goods into the United States, Korea's major overseas market. With the imposition of antidumping duties on Korean color televisions in late 1984, electronics manufacturers have been undertaking intensive development efforts aimed at introducing new models such as PAL-type color TVs (for the European market) and new products like video tape recorders (VTRs) and microwave ovens. In addition, Korean firms have established a number of overseas plants for manufacturing various consumer electronics items within the final market area in an effort to preempt trade restrictions or circumvent trade barriers.

8.12 The production shift in the direction of industrial electronics is likewise attributable in part at least to the anticipated high growth in world demand for various industrial electronics products in the future. Korea's own market for such products is also expected to grow rapidly in coming years as the country's economy continues to industrialize. Between 1987-92 for example, the domestic market for computer equipment of all types is projected to grow on average 14% per year (in constant 1982 dollars). Meanwhile, computer exports are projected to grow 22% a year on average.[3] At present telephone subscriber equipment accounts for the largest portion of both industrial electronics production and exports. Computers and computer peripherals, however, constitute a rapidly growing segment of industrial electronics in terms of both output and exports. Korean firms have been able to utilize their expertise in television and cathode ray tube (CRT) production to move into markets for various types of computer terminals for example. In 1984, total terminal exports amounted to $198.08 million, a 234.4% increase over the previous year and more than one third of total industrial electronics exports. The surge in computer-related exports in 1984 led to Korea's first trade surplus ever in computer products, with $285 million in exports against $247 million in imports. Computer and peripheral exports rose to $467 million in 1985 and are expected to reach $680 million in 1986.[4]

8.13 The strategies of Korean firms in the electronic component sector also reflect changes in the global industry. Traditionally, the primary emphasis of many Korean firms had been on the development of an in-house chip supply for their own consumer electronics products. Some had become competitive exporters of discrete transistors and relatively simple consumer ICs as well. Since the early 1980s, however, the emphasis has shifted to large scale production of more sophisticated memory ICs for sale on the open market. The rationale underlying this strategy would appear to be that, like their Japanese counterparts, the Korean electronics firms expect to be able to compete effectively in world markets for mass produced memory chips and other standardized products based on low cost manufacturing methods, given relatively low cost skilled labor and capital and low overheads. Foreign investors in

---

3/ Korea Institute for Economics and Technology.

4/ Ministry of Trade and Industry projections.

Korea's component sector, however, have thus far made no move to integrate from their assembly/test operations backward into wafer fabrication. Nevertheless, they have been investing heavily in automated assembly and test equipment in their Korean plants.

B. Electronics Industry Strategy, Outlook and Issues

Constraints on the Industry's Development

8.14 While the electronics industry has performed exceptionally over the last decade, there remain certain weaknesses in the industry's structure which need to be addressed if that performance is to be sustained in the future. One major weakness is that the industry remains principally an assembly industry based on productive, low-cost labor. While in certain areas like televisions and audio equipment production has become more highly integrated, nevertheless there has generally been little innovative product design and development activity in the past. For the most part Korean firms have concentrated on the low cost manufacture of foreign-designed products. In the past the technology needed to make such products was readily available via technology imports either in the form of licensing agreements or joint ventures, or by reverse engineering of imported products. Currently, however, it is more difficult to gain access to certain foreign technologies since their suppliers are also potential competitors with the Korean firms in major export markets. Moreover, miniaturization has also rendered reverse engineering more problematical. Hence, the restrictions (now expired) placed by Japanese video cassette recorder (VCR) manufacturers on sales by Korean licensees of their products in the US market.

8.15 Historically, investment by Korean firms in research and development activities has been low by comparison with the Japanese and US averages. In 1979 Korean electronics firms spent 1.52% of sales on R&D while Japanese firms spent roughly 5.4% and the US industry approximately 5.8%. (By 1983 the Korean ratio for electronics/electrical firms had risen to 3.1%.)[5] In the context of an increasingly competitive global industry where firms protect ever more carefully their technological know-how, the Korean electronics industry needs to be able to generate technologies indigenously to a far greater extent than in the past. This will necessitate substantial additional investments in R&D activities in the future.

8.16 Another constraint to the industry's development is its continued heavy reliance on imported components and parts, where manufacturers of consumer electronics products have had to import critical components due either to their unavailability locally or to the nonconformity of local components to the quality standards required for export production. Since component and material costs represent a substantial percentage of ex-factory costs of most electronic equipment, however, dependence on foreign supplies has reduced

---

5/ Korea Development Bank. The Japanese figures, meanwhile, rose to 7.1% in 1983. (Electronic Industries Association of Japan)

domestic value added. In addition, due to delivery delays, especially during periods of peak demand, Korean equipment manufacturers have ocassionally faced problems in meeting their own production and shipment schedules, causing them to forfeit market share. To ensure reliability of supply without having to stock large inventories of costly components, it is advantageous to maintain close geographical proximity as well as close technical links to suppliers. Furthermore, Korean electronics manufacturers are heavily dependent for their components on the same firms (often Japanese) with which they must compete in end equipment markets.

8.17 Despite the fact that productivity has been increasing rapidly in recent years, the Korean industry's performance still compares unfavorably with the industries of the advanced industrialized economies. In 1982, for example, value added per employee in Korea's electronics industry totaled $9,464 while in Japan VA per employee was roughly three times as high, or $28,779. Although wage cost differentials still more than offset productivity differentials (hourly compensation in Korea being roughly one fifth that in Japan as of 1984), Korea faces potential competition in labor intensive types of electronics operations from still lower wage countries of the region (see Table 8.6 for wage cost comparisons). In effect then, the Korean industry is under pressure from both sides. In higher value added products, it is still not able to compete on an equal footing with firms from the advanced industrial countries; in lower value added products, it may be losing its competitive edge to still lower cost producers. As suggested earlier, changes in electronics manufacturing technologies--in particular assembly automation--may alter the parameters within which the Korean industry competes but the basic challenge to become internationally competitive in higher value added, higher technology products still remains.

Table 8.6: HOURLY COMPENSATION (WAGE RATES PLUS BENEFITS) FOR PRODUCTION WORKERS IN 1984 (US $)

Asia		Americas		Europe	
Japan	6.35	US	12.59	Begium	8.87
Korea	1.36	Canada	11.51	France	7.43
Major Asian competitor	1.70	Mexico	1.70	Germany	9.57
Hong Kong	1.60	Brazil	1.23	Ireland	5.46
Singapore	2.43			Netherlands	8.60
Malaysia	0.50-1.00			Spain	4.68
Thailand	0.88			U.K.	5.87
India	0.70-1.00				
Philippines	0.39 /a				

/a Minimum rate.

Source: Electronic News, September 23, 1985.

Private Sector Strategy

8.18 The private sector consists of firms characterized by a variety of ownership arrangements: wholly foreign-owned subsidiaries, joint ventures, and wholly domestic enterprises. For the purposes of this discussion, attention is focused principally on the locally owned industry. Within that grouping, however, there is also considerable heterogeneity. While there are a growing number of large-scale enterprises with employment in excess of 1,000 persons (5.9% of total firms in 1983), the majority of firms (70%) still have fewer than 200 employees. While certain of the small-scale enterprises are independent entities which plot strategy on their own, a large number are primarily involved in supplying parts and components, subassemblies, and services to the large electronics conglomerates. In effect then their strategies are conditional upon the strategies of their principal customers. For this reason, the discussion of corporate strategies tends to be directed largely to the strategies of the five or six largest domestic electronics enterprises in Korea. Of course, those enterprises cannot be considered strictly domestic in all cases, since they or their affiliates/subsidiaries are often engaged in joint ventures, licensing agreements, or other forms of technical collaboration with foreign partners. Indeed, the nature and extent of such foreign linkages can significantly affect a particular firm's growth strategy.

8.19 Broadly speaking, three different types of private sector strategies can be identified. The major electronics firms in Korea pursue some mix of these strategies with differing degrees of emphasis. The first has been referred to frequently as the "leapfrog" strategy. Its rationale is that the nature of technological change in the electronics industry is such that firms are able to bypass certain intermediate stages of technological development and move directly to the mass production of certain high technology products. This logic is usually invoked in defense of efforts by some Korean semiconductor manufacturers to bypass intermediate levels of circuit integration in order to begin producing very large scale integrated (VLSI) circuits--in particular computer memories--in direct competition with the Japanese and, to a lesser extent, the US and Western European industries. A further elaboration of this strategy argues that dynamic RAMs (random access memories) in particular are the "process drivers" which enable firms to refine mass production techniques for eventual transfer to the fabrication of other types of devices. Thus, if a firm becomes competitive in DRAMs, it should have better prospects of penetrating other segments of the "commodity" (i.e., standardized) chip market.

8.20 A second strategy is to focus instead on moving into the intermediate range of electronics technologies in which Japanese, US and Western European firms are no longer able to maintain a technological edge and in which they might therefore be forced to concede considerable market share to Korea or other new entrants. This is an extension of the strategy which many firms have followed in consumer electronics. Indeed, to varying degrees Korean electronics firms are already pursuing such a strategy with considerable success. For example, as previously noted, Korean firms have become major exporters of "dumb" computer terminals and telephones instruments.

Korean firms have also been able to capture a sizeable share of the world market for low end oscilloscopes and digital multimeters. In the case of components, some Korean firms are highly competitive in black-and-white picture tubes and others in simple semiconductor products like discrete transistors. Yet, in integrated circuits (ICs) this strategy has not been as widely adopted. While Korean firms do process a large number of intermediate ICs (i.e., medium- and large-scale integrations, or MSIs and LSIs), they do so primarily to satisfy their own internal consumption requirements. When it comes to production for the merchant market the principal emphasis would appear to be on a "leapfrog" approach.

8.21 A third strategy is to focus on market niches rather than on high volume markets for standardized products. Such an approach requires considerable sophistication in terms of indigenous design capabilities as niche-oriented firms must be able to customize their equipment to satisfy specialized user needs, a design-intensive process. In the component sector such a strategy implies focusing on semicustom and custom IC products rather than standard memory and/or logic. At least one Korean firm (Lucky Goldstar) has pushed a variant of this strategy, linking up in a technical collaboration with a leading semicustom IC house in the US. The latter supplies the designs to the Korean partner which then fabricates the wafers in volume and ships them back to Silicon Valley for final metallization. The strategy may also involve a degree of vertical integration wherein a niche-oriented equipment manufacturer or systems vendor designs application specific ICs in-house to incorporate in the specialized equipment. Frequently this approach is suited to the design and development of complete production systems, and those firms which have large internal requirements for such systems may produce them initially for in-house use, marketing them to others only after they have demonstrated their effectiveness in their own operations. It is a human capital as opposed to a physical capital intensive process, with a substanital software component and relatively low volume hardware manufacture.

8.22 The Korean electronics industry attaches high priority to the development of its semiconductor capabilities. It has channeled enormous resources over the last few years (since 1983) in to building up its semiconductor manufacturing capacity and has plans to continue the high investment levels in integrated circuit (IC) wafer fabrication (fab) over the next few years. If for no other reason than the massive commitment of resources to this sector, the strategies of the major Korean players in the semiconductor field merit closer examination. In fact there is another important reason to examine these strategies critically, namely, the very risky nature of certain types of semiconductor activities in light of intense international competition.

8.23 As previously noted, 1983 marked the beginning of a first major round of investments in wafer fabrication facilities in Korea. Between now and 1988 the major Korean semiconductor producers (basically Samsung, Gold Star and Hyundai) are expected to invest an additional US$1.2 billion in VLSI chip production. Much of the investment capital for the first round consisted of loans from the Korea Development Bank (KDB), although some semiconductor manufacturers have also utilized international capital markets.[6] Financing will be a critical consideration in future development of Korea's semiconductor industry given the large scale of investment required on an ongoing basis if Korean firms are to keep abreast technologically. Since the firms which have invested heavily in mass memory (e.g., 64K and 256K DRAM) production are already facing severe downward price pressures and cannot expect to realize a profit on these chip sales, it remains to be seen whether outside financing will be as forthcoming for the next generation of investments (such as the one megabit DRAM). The GOK has indicated its intention to make maximum use of the National Investment Fund and new venture capital sources to finance the growth of the semiconductor industry, but neither source can provide the large sums required.

8.24 The three major semiconductor manufactureres in Korea are all pursuing the high-volume VLSI memory market to varying degrees and with slightly different product mixes. They have relied either on licensing or outright purchase of foreign chip designs. The leader in this field (Samsung) entered mass production of 64K DRAMs just as the market began its nosedive and was forced to switch some of its wafer fab capacity to other memory products. It is planning to begin volume shipments of its 256K DRAMs for the first half of 1986. This product, too, has been experiencing precipitous price declines, though there is likely to be an upward adjustment as a result of measures taken by the US Government to counteract alleged dumping by Japanese firms. The second major DRAM contendor (Hyundai) has bypassed the 64K generation and plans to move directly into 256K DRAM production. In the meantime, it has licensed technologies for certain other memory products from foreign sources and is currently producing them, though the quantities are unknown. It is handicapped by the lack of prior experience in the semiconductor business, but it remains to be seen how serious that handicap will prove to be.[7] The third major chip manufacturer (Lucky Goldstar) has adopted a more cautious approach to its semiconductor investment strategy. This may be partially a function of its joint venture with a major US telecommunications and semiconductor firm and partially a function of the fact that its chip production is geared principally to the domestic rather than the export market. Where it does export, it has chosen to rely principally on a "buy back" strategy with its joint venture partner or with other technology licensors (e.g., the gate array manufacturing agreements alluded to above). At the same time, it is pursuing

---

6/ Recently, Samsung, the leading semiconductor manufacturer in Korea, issued convertible bonds worth US$20 million in the Eurobond market. While the capital raised was intended primarily for expansion of consumer electronics production facilities abroad, this represents a departure for Korean electronics firms in that it is the first instance in which foreigners have been permitted to invest directly in Korea's stock market.

7/ Already it has encountered serious technical and cost problems in its Silicon Valley-based production line, which it has been forced to shut down as a result.

R&D on one megabit DRAMS and may be one of the first Korean firms to have a mass production capability in the next generation of memory chips.

8.25 In computers and peripherals Korean firms are slightly less advanced on the whole than in semiconductors. Until very recently the industry has consisted largely of two sets of activities: local firms producing computer peripherals like monitors for export on an OEM basis; local firms producing microcomputers to supply a protected domestic market. In the case of the latter, the GOK has also supported local computer manufacture through its procurement policy, for example, by ordering 5,000 personal computers in 1982 from 5 local firms for dissemination to educational institutions. More recently, the GOK has contracted with several local firms to supply superminicomputers for incorporation in a planned nationwide government computer network for handling the information processing needs of the various municipal and lower level administrative units.

8.26 In the last two years significant changes have occurred in the Korean computer and peripheral industry. First, the GOK's attitude toward foreign direct investment in this sector has become somewhat more positve, with the result that a number of foreign computer firms have invested in local production--in some cases through wholly owned subsidiaries but more commonly through joint ventures--to gain access to the Korean market as well as to provide a base for regional exports and for local sourcing of components and peripherals to supply their global operations. While the domestic market for personal computers is still protected, the GOK is committed to the liberalization of imports of small- and medium-size computers by 1988. Meanwhile, local firms have met with growing success in exporting fully configured personal computers, in particular to the US market.

C. Public Sector Role

The System of Incentives and Government Attitudes

8.27 Government policy has played a critical role in the electronics industry's development since 1969 when the Electronics Industry Promotion Law was promulgated. That law designated electronics a strategic export industry, which made it eligible for a number of special incentives. Specific measures to promote the sector's development were incorporated in the Eight-year Electronics Industry Development Plan (1969-1976) and the successive Five-year Economic Development Plans. In 1981 the Government revised the Electronics Industry Promotion Law and drew up an executive plan for making electronics a highly advanced industry in Korea. Among the proposals in that plan was the establishment of an Electronics Support Fund, to be financed by public and private sector contributions. The size of the ESF was to be W 7.5 billion and the money was to be lent at preferential rates to firms investing in high priority areas. In particular, firms establishing R&D subsidiaries overseas were to be given preferential access to loans. This fund was recently abolished, as Government has shifted in focus to functional incentives.

8.28 In view of the fact that the electronics industry in Korea has developed principally as an export industry, it has also benefited from the basic export incentives which were instituted beginning in the early 1960s. The systems of incentives are of two types: one granting duty-free and unrestricted access to imported intermediate inputs needed for export production, including for firms which are merely suppliers to export manufacturers; the other granting automatic access to bank loans for working capital requirements associated with export activity. Besides those incentives and other measures intended to foster export competitiveness in general, the electronics industry has been the target of a number of other promotional policies relating to trade and tariff, procurement, patent and licensing, fiscal and financial incentives, and foreign investment.

8.29 Trade and Tariff Policy. As of 1984 there were some 185 electrical and electronics goods whose import into Korea was restricted (37% of all electrical/electronics products listed). Most of the goods on the restricted list are considered infant industry products requiring temporary protection. Under the import liberalization plan all electronics items currently restricted will be freely importable by 1988. The move to liberalize electronic imports has resulted in part from mounting pressure from Korea's major trading partners to open up its own markets as a condition for ensuring entry of its products into foreign markets. While electronics firms--as well as other strategic industries--no longer enjoy the special tariff preferences they once did, there are certain generic preferences which have particular relevance to electronics firms. For example, firms pay a lower tariff rate on equipment which is imported for R&D purposes. Since electronics is increasingly R&D intensive in Korea, this provision is of special benefit to firms in this sector.

8.30 Procurement Policy. The role of government procurement policy in fostering the computer industry's development is important, particularly for the telecommunications equipment industry. The GOK has supported the joint development by the Korea Telecommunications Authority's (KTA) research arm, the Electronics and Telecommunications Research Institute (ETRI), and four private sector firms of an indigenous electronic switching system for eventual incorporation in the domestic public telecommunications network. It has licensed those firms to produce the switching equipment for sale to the KTA. Such procurement practices also constitute indirect support for the component industry, in particular the IC industry which is a major supplier to the telecommunications equipment as well as the computer industry. As purchaser, the Government will need to exercise caution so that foreign firms can begin to compete in the Korean market.

8.31 Patent and Licensing Policy. The protection of intellectual property rights (IPR) has become a controversial issue within the Korean electronics industry. While products patents have never represented a particularly effective means of protecting technology from duplication in the case of electronics products, the issue of protection for computer software is especially sensitive to would-be suppliers of technology to Korea. It is also sensitive to the Korean software industry, which has relied extensively on translation or simple modification of imported software in the past. While recognizing the contribution that protection of software property rights would make to the acceleration of technological development in Korea's computer industry (e.g., by inducing Korean firms to do more innovative software design), Korean entrepreneurs are nevertheless reluctant to endorse such protection because of the potential diminution in their profits on software sales that might result from increased royalty payments abroad. While not presently a major issue, copyright protection for integrated circuit designs may well become a critical concern as the Korean semiconductor industry develops. Such protection is apt to be a concern not simply to foreign semiconductor firms but to Korean companies themselves which undertake their own design work and stand to lose potential revenues to imitating rivals.

8.32 Tax Incentives. Since 1982 incentives to promote the development of specific industries have been substantially reduced. Nevertheless, special incentives continue in effect for six strategic industries, among them electronics. This industry (along with the industrial machinery sector) has a choice between two options for fiscal relief for facilities investments: tax credits of 3-5% of the amount invested or accelerated depreciation (other strategic industries have only the latter incentive available). Besides these limited incentives targeted specifically at electronics, the industry also benefits from a number of functional incentives, for example, technology and manpower development. One of the most important incentives is the allowance of firms to set aside a percentage of profits in a reserve fund (exempt from taxation for a fixed time period) for eventual investment in R&D activities. In the Sixth Five Year Plan the GOK has chosen to rely more extensively on such activity-specific--as opposed to industry-specific--incentives to promote technology development and manpower training.

8.33 Financial Policy. The GOK provides a number of financial inducements to investment in the electronics industry. First, it requires that commercial and state-owned banks reserve a certain portion of their loanable funds for R&D lending, a fact which in the context of persistently tight credit conditions consitutes an effective incentive. Furthermore, the Government has instituted measures in the last several years designed to stimulate technology related lending and venture capital activity. It was instrumental, for example, in the establishment of the Korea Technology Development Corporation (KTDC), which lends to R&D projects as well as taking equity positions in technology-oriented start-up firms (though its equity investment portfolio remains fairly limited). Approximately one third of KTDC's portfolio projects are in the electronics field. More recently, the GOK has adopted a new law for the promotion of venture capital activities with a view to encouraging greater risk-taking activity among Korean financial intermediaries. While it is not yet known how effective the law will be, to the extent that it does encourage greater entrepreneurship in the manufacturing sector, it should stimulate new investment--especially start-up investment--in electronics.

8.34 The Government's National Investment Fund (NIF) has played a relatively small role in the development of electronics compared to other strategic industries. From the Fund's establishment in 1974 through 1982, NIF loans were made available to strategic industries at preferential rates; however, since 1982 preferential interest rates on NIF loans were essentially abolished. Presently, electronics is one of the major recipients of NIF loans, but the total size of the NIF is about $600 million and the GOK intends to continue to reduce its size. Between 1976 and 1986, NIF funds lent to the electronics industry amounted to roughly 4.2% of the industry's total capital investment.

8.35 Through collaborative research projects between government-affiliated research institutes and private sector firms the GOK has also contributed financial as well as technical resources to electronics R&D efforts. The Electronics and Telecommunications Research Institute (ETRI) is involved in several collaborative projects with the private sector, including public switching systems, VLSI design and process technologies and advanced computer system architecturing and design. The GOK allocates budgetary funds each year to the financing of certain high priority "national projects," many of which involve collaboration between sector R&D labs and public research institutes.[8/] Another nonfinancial incentive which the GOK grants to stimulate R&D activities is the exemption of R&D personnel from military service.

8.36 Foreign Investment Policies. Korea has historically sought to limit wholly foreign owned direct investment in its industries. To the extent that foreign direct investment has been permitted, it has usually been in the context of majority Korean owned or 50:50 joint ventures (see Table 8.7). Of all electronics-related foreign invested projects between 1962 and 1983, 63% were in those two categories. Those projects, however, represented only 30% of the total equity in foreign invested projects. Wholly foreign-owned projects on the other hand, while only a quarter of the total number of projects, accounted for 61% of the total equity invested.

8.37 In recent years, the GOK has liberalized considerably its foreign investment law. The Revised Foreign Capital Inducement Law of December 1983 is designed to encourage greater foreign direct investment (FDI) through the streamlining of approval procedures and the reduction of restricted areas. Once again this has special significance for electronics, given the importance of FDI as a means of technology transfer in this sector. The most significant change was from a positive to a negative list. Almost all restrictions on capital and profit repatriation have been removed under the revised law. Certain restrictions remain on foreign investors, however. For example, they are not eligible for subsidized credit unless they are in a high-priority industry or are part of a majority Korean-owned joint venture (of course high technology electronics is a high priority area). Moreover, in the electronics field, 100% foreign owned manufacturers are required to export at least 50% of

---

8/ Here again, due to the strategic significance attached to electronics, this industry is the recipient of substantial government support through various national R&D projects. In 1983 some 182 research projects of 131 industrial firms were selected as national R&D projects and about US$28 million was contributed by the GOK to these projects. In addition, the GOK is supporting seven special projects in semiconductors and bio-engineering with funding of W 35.7 billion (approx. US$40 million).

Table 8.7: ELECTRICAL AND ELECTRONICS: FOREIGN EQUITY INVESTMENT (APPROVALS) BY SHARE OF FOREIGN OWNERSHIP, AS OF AUGUST 1983 (cumulative from 1962)

	Value (US$ million)	Percentage of total	No. of new projects
1-49%	63.8	24.3	58
50%	15.8	6.0	62
51-99%	22.0	8.4	23
100%	161.3	61.4	48
Total	262.9	100.0	191

Source: Ministry of Finance.

their output. (Most joint ventures on the other hand can sell all of their productin domestically.) Despite the remaining restrictions, the liberalization has led to an upsurge in foreign direct investment in Korea's electronics industry in recent years.[9] (Of course, other factors have contributed to the expansion of FDI as well, such as the growth in Korea's domestic market and its low manufacturing costs.)

9/ Among the major foreign investments in Korea's electronics industry in the last few years are:

- Oriental Telecommunications Company (OTELCO), a 50:50 joint venture between Ericsson (Sweden) and Oriental Precision Company, established in November 1983 to supply digital switching equipment to the Korean Telecommunications Authority;
- Fujitsu Ltd. (Japan), which plans to build a factory to make personal computers;
- IBM (USA), which gained permission in early 1985 to establish IBM Korea Systems Inc. to manufacture and sell IBM personal computers;
- Hewlett-Packard (USA), which has a joint venture with Samsung to assemble the HP 3000 minicomputer and an "Asian personal computer," established in June 1984;
- Gold Star Fiber Optics, a 50:50 joint venture between Lucky-Goldstar and AT&T (USA), established in January 1984;
- Gold Star-Honeywell, a 50:50 joint venture between Lucky-Goldstar and Honeywell (USA), established in November 1983.

Industry Prospects

8.38 Access to Capital. As the electronics industry has evolved, many segments of it have become more capital intensive. This has been the result of increasing competitive pressures to automate production as well as the increasing sophistication of the production equipment required to make ever more complex electronics products. As minimum capital requirements for entry into particular types of electronics production have risen, the importance of access to capital at competitive rates has increased correspondingly. On the one hand, Korea's electronics industry has been compelled by the need to adjust to rising wage levels to introduce automated equipment into existing operations. (It should be noted, however, that wage levels have stabilized in recent years and the electronics industry still remains relatively labor-intensive.) On the other hand, it has chosen to diversify heavily into semiconductor processing, an operation whose capital intensity has been rising steeply over the last decade. The latter development in particular has thus made financing considerations increasingly central to the health of Korea's electronics industry.

8.39 The Electronics Industry is a High-Risk Business. The risks involved are of both the commercial and technological variety. In the past, Korean electronics firms have tended to avoid the latter risk to a large degree by concentrating their activities on the replication and mass production of products which have already passed through the initial stages of the product life cycle and have thus been tested in the market. As Korean firms restructure their activities toward more technology-intensive products and processes, increased exposure to technological risk becomes inevitable. Korean firms are now being forced to increase spending on R&D, design engineering, prototype production and testing, and other precommercialization activities and as this happens, the nature and level of risk changes. The financing of investments in such high risk activities may prove problematical within the existing capital market structure. In the Korean context, the difficulties stem from the overall inadequate capitalization of the very large firms and the inability of commercial banks to either adequately assess the riskiness of loans or to change a sufficiently high risk premium if they considered it sound bank practice to do so. See Chapter 5 on the immaturity of the Korean capital market.

8.40 Policymakers have perceived the need for public sector involvement in this area. That involvement has taken a variety of forms. For example, the Korean Technology Advancement Corporation (K-TAC) has served as a source of investment capital for the development to the commercial stage of new products/processes generated by the research of the Korean Advanced Institute of Science and Technology (KAIST) and other government research institutes. Similarly, with Bank support, the Korean Technology Development Corporation (KTDC) has made funds available to the private sector for investment in specific R&D, technology acquisition and other precommercial investment projects. Recently KTDC has expanded its operations to include venture capital-type equity and other risk sharing investments in certain high technology enterprises. Moreover, the GOK is actively encouraging other financial intermediaries to undertake venture capital operations, especially in support of technology intensive projects in electronics and other fields.

8.41 Human Capital. The evolution of the industry has altered the structure of demand for labor towards higher-skill categories. Technicians and engineers are in especially great demand. The changing composition of the electronics labor force is the result of a combination of the increasing automation of electronics production on the one hand and the relative increase in nonproduction activities (e.g., R&D, design, engineering and marketing) on the other. As automation occurs, the number of direct operators needed to produce a given level of output diminishes while the number of engineers and technicians needed to supervise, service and maintain the automated equipment increases. Thus the Korean electronics industry has come to the point where the availability of skilled scientific, engineering and technical labor represents the single most critical potential constraint to its future development. Indeed the industry currently suffers from a shortage of high level engineers experienced in R&D activities.

8.42 In human cpaital investments, Government performs a crucial function. Educational institution building is necessarily a long-term process, however. In the interim, there are a variety of ways by which the Government as well as the industry itself can seek to prevent the emergence of serious bottlenecks in the supply of certain categories of skilled labor. One widely employed method is the granting of scholarships for foreign study in high priority disciplines. Another is the hiring of foreign technical consultants in fields where Korean expertise is lacking. Still another widely used by Korea is the establishment of foreign R&D and design facilities to tap into overseas supplies of skilled engineering and technical labor.[10/]

8.43 Access to Foreign Technology. Somewhat paradoxically, as Korea's electronics industry develops technologically, the need for access to foreign technology is apt to increase rather than decrease; Korean firms are no longer able to rely on proven product and process technologies, but rather must begin

---

10/ Korean firms have had mixed results with the last strategy. For there may be constraints to hiring the highest quality personnel in the United States. Certain employers are reluctant to hire professionals who have gone to work for foreign firms. Moreover, the most talented engineers generally expect stock options as part of their benefit package, something that closely held Korean jaebol generally are unable to offer. At the technician level the time horizon for building up effective local training institutions may be somewhat shorter. The scope for private sector involvement may also be greater. To a greater extent than is the case with engineers, the training of technicians occurs on the job rather than in the classroom. Nevertheless, the Government may still perform a valuable role in building up technical training institutes which, while responsive to the needs of the private sector, do not suffer from the deficiency of some in-house training programs, viz., that the skills imparted are so firm specific that they defy transfer across firms if and when a skilled employee chooses to change employers. The Government may also have greater incentive to invest in the development of skills which are not currently in great demand but are expected to experience growing demand in the future.

to master more sophisticated, less mature production processes and product designs if they are to remain internationally competitive. They are not, however, in a position to generate all the necessary products and processes based on their own indigenous R&D and design capabilities. Rather, they must be able to acquire the most advanced product and process technologies abroad so as to be able to assimilate and, where appropriate, adapt them to local industry requirements. While licensing agreements served Japanese electronics firms well during their period of accelerated learning as an avenue for acquiring the basic technologies they first reproduced and later refined, those agreements have diminished usefulness in the context of the present day industry. This follows because the major international electronics firms--including, or perhaps especially, the Japanese--are increasingly reluctant to license advanced technologies to potential competitors. On the other hand, the intense pressure which Japanese firms have exerted on their foreign competitors in the semiconductor markets, for example, has enabled Korean firms to license or purchase foreign technologies which would not have been available to them under different circumstances.

8.44 Moreover, there may be a symbiotic relationship between technology transfer from abroad and indigenous technological development. The Korean industry is in a better position to bargain for favorable terms in technology licensing or other transfer agreements inasmuch as foreign technology suppliers have reason to be convinced that the Korean industry is close to being able to generate the technology on its own. Moreover, technology suppliers are apt to be more forthcoming with proprietary technological know-how when the buyer/licensee has some technological expertise of his own from which the supplier may derive benefits.[11/] Indeed, amongst developed country electronics firms cross licensing and technology exchange agreements are increasingly common.

Effects of Market Structure

8.45 While there were some 700 electronics firms in Korea as of 1983, the three largest account for approximately 40% of electronics exports and production. The high degree of concentration in this and other industries has given rise to a policy debate regarding an appropriate market structure which is both economically efficient and socially beneficial. In the case of the electronics industry, there are both benefits and costs associated with the existing market structure. The large size of the leading Korean electronics firms has enabled them to reap economies of scale as well as giving them sizeable financial resources from which to fund the increasingly large investments required to remain competitive in the electronics business. Their diversified product portfolios has also permitted a degree of cross-

---

11/ The willingness of a leading US semiconductor manufacturer to grant a second-source license to a Korean firm for one of its popular microprocessor lines, for example, was contingent on the latter's ability to meet the former's stringent quality and reliability requirements. At the same time, it has now licensed its most advanced microprocessor line to the Korean firm.

subsidization of investments in higher risk projects with low or negative margins from the revenues earned on established product lines with high profit margins. In addition, their large asset bases have given them privileged access to capital markets to raise funds for major new projects like the recent wave of VLSI wafer fab investments.

8.46 Many of the small- and medium-scale enterprises in Korea are suppliers to the large conglomerates. In certain instances they may be direct competitors with the latter but normally SMIs are more specialized whereas the conglomerates are mass producers of a wide range of products. The absolute size of small- and medium-scale producers has been increasing steadily, suggesting that those firms are able not only to survive but to grow in a market dominated by a half dozen large scale enterprises. They have grown, moreover, in an environment in which those supplying the export market (whether directly or indirectly) have had to meet international quality standards at competitive prices.

8.47 Nevertheless, there remain certain weaknesses in the infrastructure of supplier and support industries in Korea. These weaknesses have been highlighted by the rapid expansion of semiconductor manufacturing capacity in recent years. The materials and other suppliers of inputs for wafer fabrication have often lagged behind their customers technologically; in some cases--e.g., certain electronics grade chemicals, lead frames, precision tools and dies, etc.--few if any local suppliers have been able to meet the precise specifications of semiconductor firms. Thus, many direct production materials have continued to be imported. Another potential problem area in Korea is the tendency for the electronics conglomerates to internalize much part and component production which could be undertaken more efficiently by specialized SMIs. The large electronics conglomerates maintain networks of dedicated suppliers while at the same time sourcing on a competitive basis from many others. The former group have the benefit of their single customer's technical assistance, quality control, training and financial quarantees on bank loans. The independent suppliers, however, are in a more precarious position and may require bolstering if they are to overcome their technical and financial constraints.

8.48 R&D Issues. If the electronics industry is to become and remain competitive in sophisticated products like VLSI memory ICs and high performance mini- and microcomputers, it will need to augment considerably its R&D activities. Already that process has begun. For example, the number of research laboratories in electronics increased from 11 in 1980 to 28 in 1983. Similarly, the research staff, comprised of college graduates and above in the electrical and electronics industries, doubled in the 2 years from 1981 to 1983 from about 700 to 1,400. Presently the major electronics firms in Korea spend roughly 4% of sales revenues on R&D investments. They have plans to raise that by several percentage points in the next few years to bring their R&D:sales ratios into line with those of their major developed country competitors. Still, given their smaller revenues, the absolute amounts spent on R&D are likely to continue to remain well below those of many developed country electronics firms. (Indeed, in 1982 total Korean R&D expenditures for all purposes were smaller than the individual R&D budgets of each of the three largest U.S. computer manufacturers.)

8.49 Public Sector's Role in Other Countries. The Korean Government has drawn lessons in the past from the experiences of other countries in the promotion of their electronics industries. Presumably it can continue to learn from those experiences in the form of both positive and negative lessons. Korea is neither the most interventionist nor the least interventionist in terms of public policy toward electronics. In some countries public sector corporations (or parastatal enterprises) are the main producers of electronics hardware. In others the government has adopted an almost purely laissez-faire stance vis-a-vis this and other industries.

8.50 Among developing countries, India historically and Brazil currently are examples of significant state intervention in electronics. China likewise fits into this category. Like India, China has been involved in a process of reassessment of the degree of direct government involvement in electronics, though in the two countries the terms of the discussion differ considerably. Both have begun to liberalize, in particular, government restrictions on foreign participation in their industries. In addition, India at least has been reviewing procurement policies with a view to stimulating more private sector involvement in supplying telecommunications equipment and components as well as other electronic equipment to the public sector. It has removed limitations on production capacity for most types of electronic equipment and components. Procedures for import of raw materials needed in electronics manufacture have also been streamlined.

8.51 Brazil has followed a markedly different course in recent years, extending rather than limiting government intervention in the electronics industry, in particular in computer related activities. The main thrust of its policies has been to create a protected market within which the domestic private sector can earn rents by restricting foreign participation in a broad range of market segments. The public sector is not itself a major producer in Brazil as it is in China and India, so it is not a potential competitor with the domestic private sector. By reserving the local market for domestic producers Brazil expects to stimulate the development of indigenous technology over the long run.[12/]

12/ In the short run there may be substantial costs inasmuch as local computer users must pay higher prices for equipment whose performance oftentimes does not match that of foreign equipment. Whether the cost is justifiable depends on the size of the price differential for machines of equivalent performance and the length of the learning period needed to reduce costs to levels comparable to those of imports. Limited evidence suggests that for low-end microcomputers at least, Brazilian firms have been able to lower prices to a relatively small margin above their foreign equivalents in a timeframe of two to three years. For medium manage systems (e.g., IBM-PC compatible systems) the price differential has remained relatively high. The classic form of import-substitution is based in part on the size of the domestic market, but it involves clear costs to domestic consumers and risks retaliating trade actions. See, e.g., Claudio Frischtak, "The Informatics Sector in Brazil: Policies, Institutions and the Performance of the Computer Industry," Paper prepared for National Science Foundation Symposium, August 1985, pp. 28-31.

8.52 Countries lying at the opposite end of the intervention spectrum include Hong Kong and, to a somewhat lesser degree, Singapore. The governments of those two city-states have maintained their economies as free ports with unrestricted trade and investment flows. This has led to a massive inflow of foreign capital into their electronics industries, which have been almost exclusively export-oriented. The result of this laissez-faire approach has been rapid growth in electronics output and exports, but with the locally owned industry lagging behind and generally dependent upon foreign-owned ventures. (At the same time, both industries have been highly vulnerable to fluctuations in the world market for specific electronic products.) The level of indigenous technological development remains low in both locations, though in recent years the Singapore Government has adopted a more interventionist stance in an effort to strengthen the indigenous skill and technology base.[13/] The far more liberal approach of Hong Kong has retarded the process of technological development to a degree, inasmuch as investments have been made with a view primarily to short-term profitability. The Hong Kong Government has sought, however, to promote the application of microelectronics-based technologies to certain traditional industries like textiles and garments.

8.53 Future Directions. There is no doubt that the Government has sought explicitly to encourage the development of high technology industries like computers and semiconductors by designating them "strategic industries" entitled to certain preferential treatment. In effect the singling out of these two subsectors of the electronics industry for special status constitutes a policy to reconfigure the electronics industry as such away from low value-added, low-technology products and activities to high-skill, high-value added ones. While the extent of direct Government intervention in support of the electronics industry is relatively limited in Korea, the combination of support measures and activities directed at technology and R&D intensive activities in general and electronics in particular does suggest that Korean industrial policy is functionally nonneutral with respect to this sector. The priority that electronics projects receive among national R&D projects discussed above, the Government's budgeting support for public R&D institutes in this field, and its use of procurement policy to encourage development of certain production technologies all corroborate the government's efforts to foster the emergence of a technologically more advanced and relatively were autonomous electronics industry in Korea.

13/ The Government's initiatives have been directed primarily at altering the market signals to which private investors--both local and foreign--respond. For example, it initiated a sharp increase in wage rates in the early 1980s in an effort to induce firms to invest in labor-saving technology and higher value added activities. (It has partially reversed that policy in the wake of recent economic difficulties and the slump in the electronics sector in particular.) It has also offered generous incentives (e.g., interest free loans) to firms willing to invest in specific types of production (e.g., IC wafer fabrication).

8.54 The rationale given for government intervention to promote the electronics industry is that, in the context of Korea's export-led growth strategy, the Government must anticipate changes in the country's comparative advantage in order to smooth the adjustment process which will prove necessary to sustain export competitiveness. More specifically, it is argued, Korea's comparative advantage in labor-intensive assembly activities is likely to be eroded as wage levels rise with the general standard of living. In effect Korea is seeking to create a comparative advantage in those activities which are intensive in human capital, i.e., in skilled labor.[14/] In the case of government support for R&D intensive activities, it has been argued that, due to market failures, there is a tendency for private firms to underinvest in such activities.[15/] While the approach may sound reminiscent of the HCI experience of the 1970s where Korea opted to develop its capital-intensive industries, these are both more sound reasons for some functional industrial interventions in the areas of human capital and R&D as well as greater experience on the part of policymakers in terms of avoiding heavy-handed industry-specific or product-specific public involvement. The major dilemma which will face Government will be whether to underwirte losses if the industry either fails in its gamble or succeeds but fails to earn sufficient profits.

8.55 There is little doubt the industry must continue to upgrade itself technologically. Questions abound, however, on the limits of that development. The first question is whether Korea can realistically expect to acquire competitive advantage in high technology segments of the industry. It is plausible to assume that, if the Korean electronics industry were to continue

---

14/ While unskilled labor supply is a simple function of demographic trends and labor force participation rates, the supply of skilled labor is created through education, training and experience. In the first area the primacy of the Government's role is hardly subject to question. In the area of training there would appear to be ample scope for both public and private initiatives, while in terms of experience the private sector will almost invariably be the principal supplier. The issue for the Government then becomes how to induce the private sector to create the kinds of jobs which will provide the requisite experience to add to the skill pool of the labor force.

15/ This is attributed to the fact that the social benefits exceed the benefits accruing to the private firms. In other words, the individual firm weighing the private gain from R&D investment against its cost would choose a lower level of such investment than would the public at large, assuming the public possessed adequate information about both costs and social benefits. The reason for this divergence between private and public gain is the fact that R&D enjoys externalities inasmuch as the technical know-how generated cannot be contained effectively within the individual firm but inevitably diffuses to other firms. It contributes to the reservoir of technical expertise and skills from which other firms and the society at large stand to gain. Hence, the rationale for an active government role in encouraging R&D activities.

producing the same products with the same techniques, it would eventually be undercut by lower cost competitors. The industry no doubt recognizes this eventuality and would take certain measures independently of government encouragement to prepare for it. Of course, one option open to the industry is to phase down operations and shift resources to other industries. It is not at all obvious, however, which industries could fill the void left by electronics. Moreover, the firms active in the electronics industry have accumulated considerable expertise and experience which would not be readily applicable to other industries. While the industry could postpone the loss of competitiveness perhaps by successive upgradings of technology to diminish the importance of labor and other relatively costly inputs in existing product lines, the prospects of the industry would continue to be linked to one of the slower-growth markets, viz., consumer electronics. Thus, if the Koreans are to remain in the electronics business, they are faced with formidable pressures to reorient their operations toward more skill- and technology-intensive sectors like computers and semiconductor wafer processing. To say that is not to answer the question of whether they can survive in those markets in the face of developed-country competition.

8.56 If they are to face that competition however, they may in the short run at least find themselves at a competitive disadvantage by virtue of their limited experience and narrow technology base. Can that disadvantage be remedied in an acceptable time frame and at acceptable cost? These are the questions the Government may ultimately have to face in deciding whether and to what extent to intervene in the electronics industry. Presumably the Government's intervention would have as one of its primary objectives the acceleration of the process of technological learning by private sector firms. In the advanced industrialized countries government intervention in the form of support for R&D, procurement, and implicit production subsidies has proven instrumental in accelerating the process of technological learning in electronics. In effect, it has partly socialized the enormous risks of many state-of-the-art technology development projects while to an extent also bringing the research results into the public domain (at least with regard to precommercial research). Based on the effectiveness of such intervention in many developed countries, one might expect a similar sort of government involvement to be considered in Korea.

8.57 At the same time one needs to bear in mind certain differences between the developed country experience and that of Korea and other NICs. First, government support for electronics R&D is building on a much firmer technological base in the US, Japan and Western Europe.[16/] Second, there is a major difference between Korea and the advanced industrial economies in the

---

16/ The latter all have well developed infrastructures of research establishments--whether in the universities, government laboratories, or private sector firms--pursuing basic and applied research in a number of fields (e.g., metallurgy, solid state physics, materials science, chemistry, etc.) with important implications for technological development in electronics. Korea has similar institutions (e.g., in KAIST--the Korea Advanced Institute of Science and Technology), but the scientific research tradition is much shorter lived. Much so-called R&D performed in Korea continues to be of the product modification/adaptation variety.

financial resources at the disposal of the public sector to support R&D and government needs to be very selective in its choice of project support with public monies not to merely subsidize private efforts.[17] As a general rule, intervention should be limited to those activities where there are significant externalities and should be geared to the extent possible to stimulate private sector initiative. Third, since Korean firms are much more poorly capitalized than say US firms, Government must be careful to avoid becoming a risk partner in the high-tech electronics field. A major difference in risk-bearing exists between Korea and its more developed competitors, inasmuch as Korea's capital (and in particular equity) markets are still immature. Therefore, Government must clearly signal to the industry that moral hazard behavior will be penalized. This may of course prove to be very difficult inasmuch as the government has certain industrial objectives in mind.

D. Concluding Issues

8.58 The outcome of Korea's high-risk electronics strategy remains to be determined. In semiconductors the major Korean producers began gearing up production just as the world market went into its worst slump in history. The Korean firms initiated their investments in 64K DRAM technology with the expectation that they would not be profitable but would provide valuable learning experience for rapidly moving into still more sophisticated memory markets (e.g., the 256K and one megabit DRAM). Even there the prospects for profitability are small. While they were able to borrow heavily to finance the first and second rounds of VLSI processing investments, it may prove far more problematical to raise the substantial sums needed for the next round (the one megabit DRAM), especially if the initial investments are not generating profits. Thus, financing considerations loom on the horizon as a major issue which will confront the government if it is committed to the long-term viability of Korea's semiconductor industry strategy. It must weigh the potential costs to the economy of the strategy's failure against the cost of future intervention. The consequences of failure might be sufficiently severe that the Government would choose actively to dissuade private firms from pursuing such a high risk strategy, perhaps by informing the private sector of the Government's determination not to rescue firms which may encounter

---

17/ In the first instance, the revenue base of the GOK is relatively small compared with that of the US or Japan. Moreover, the magnitude of expenditures which would be required to close the technological gap separating Korea from the latter two countries in basic and applied research in electronics would strain even a much larger government budget, given the competing demands for public sector resources. For this reason, it is imperative that, to the extent that the GOK should choose to support R&D activities, it be highly selective in its choice of projects in an effort to maximize the effectiveness of its expenditures and ensure that its resources are not too thinly spread over a wide range of R&D projects with the result that none receives sufficient support to achieve a critical mass of R&D effort and expertise.

financial difficulties as a result of their unsuccessful pursuit of this high risk strategy.

8.59 A second issue which policymakers in Korea are already being forced to confront, but which should become more pressing in the future, is how to effect the transition to a more skill and technology intensive electronics industry while minimizing one adjustment burden on the current electronics workforce. As the industry upgrades technologically, employment opportunities for the semiskilled production will be rendered superfluous by automation. The Government may need to consider the promotion of retraining programs to enable existing workers to upgrade their skills so as to qualify, for example, for the growing number oftechnician positions. The Government will also need to keep abreast of trends in the electronics industry to be able to anticipate changes in the skill composition of labor demand over time. Without sufficient foresight in manpower planning, the Korean industry could be handicapped in its efforts to exploit new technological developments and new market opportunities.

8.60 Over the long term, if the Korean electronics industry is to maintain its dynamism, greater attention will need to be given to the strengthening of the small- and medium-scale component suppliers and service industries supporting the activities of the large conglomerates. Of course, large Korean firms cannot be expected, on a long-term basis, to purchase inputs from local component suppliers if the latter cannot meet price, quality and reliability requirements of internationally competitive production. Thus, it may be necessary to assist the upgrading of capabilities of component suppliers as a precondition for promoting local sourcing. In the case of those components where scale economies obtain, ther may be a need for lenders to regulate entry to ensure that the market does not become overcrowded with inefficient producers. As the market expands, new firms can be permitted to build up capacity.[18/] Perhaps certain of the small and medium scale component suppliers, given appropriate incentives, could become major international competitors in their own right.

8.61 Pressures for protectionism in Korea's major export markets could very well intensify if Korea succeeds in its electronics strategy. Already the Japanese semiconductor industry, by virtue of its exceptional effectiveness in capturing VLSI memory markets, has been confronted with strong pressures from of the US semiconductor industry. The Korean electronics industry has already confronted such protectionism in consumer electronics and should be prepared to face it, perhaps with far greater intensity, in areas like

---

18/ An additional problem confronting the local component industry which may also argue in favor of a controlled entry policy is that many foreign electronics investors are reluctant to do business with very small enterprises due to concerns about reliability and adequacy of supply. A proliferation of small scale enterprises might render the development of strong linkages between local component suppliers and foreign-invested electronics equipment manufacturers problematical.

semiconductors and computers.[19/] Whatever the economic rationale for continued tariff protection of certain segments of the electronics industry, Korea will most probably be confronted with the necessity of gradually reducing tariff levels on most if not all electronics products if it is not to face possible retaliation by its major trading partners in the electronics field, in particular the United States.

8.62 There are relatively few countries from which Korea can derive useful lessons in this particular phase of its industry's development in the sense that few countries have set such ambitious goals for their electronics industries' development and even fewer have tried to compress that development into such a short time span. Indeed, if Korea should succeed, it will no doubt have valuable lessons to offer other countries.

19/ Korean electronics firms like to stress the commonality of interest between themselves and US electronics firms in devising methods of sucessfully competing with the Japanese industry. To this end they suggest that US firms combine their design expertise with the low cost mass production techniques of the Korean industry. While there have been a number of licensing agreements and joint ventures between US and Korean firms in recent years, the fact remains that as a whole US electronics firms are also concerned about the prospect of Korean competition in the long term, especially in such areas as semiconductors.

REFERENCES
(for Volumes I and II combined)

Adams, Gerald, and Lawrence Klein, 1983, eds. Industrial Policies for Growth and Competitiveness: An Economic Perspective, Lexington, Mass.

Ashoff, G., 1983, "The Textile Policy of the European Community Towards the Mediterranean Countries: Effects and Future Options", Journal of Common Market Studies, September.

Balassa, Bela, et al., 1982. Development Strategies for Semi-Industrial Countries. Baltimore: Johns Hopkins University Press.

Balassa, Bela, 1985. "Exports, Policy Choices and Economic Growth in Developing Countries after the 1973 Oil Shocks," Journal of Development Economics.

Bank of Korea, December 1985, "Input-Output Structure of the Korean Economy: 1983", Quarterly Economic Review.

Bank of Korea, 1985, Revised Macroeconomic Model for the Korean Economy, Quarterly Economic Review, Bank of Korea, pp. 19-35.

Bhagwati, Jagdish, 1978. "Foreign Trade Regimes and Economic Development: Anatomy and Consequences of Exchange Control Regimes," Cambridge, Mass.

Bhagwati, Jagdish and T.N. Srinivasan, 1979. "Trade Policy and Development," Dornbusch and Frenkel, eds., International Economic Policy: Theory and Evidence, Johns Hopkins University Press.

Brander, J. and A. Dixit, 1982. "Tariff Protection and Imperfect Competition," in Kievzkowski, ed., Monopolistic Competition in International Trade, Oxford University Press

Brander, J. and B. Spencer, 1983, "International R&D Rivalry and Industrial Strategy," Review of Economic Studies.

Bruno, Michael, 1985. "Introduction" in World Development.

Byaumgisha, F.K. (1984), The Textile Industry in Korea: A Background Report, Mimeo, World Bank, Washington, DC.

Cable, V. and B. Baker, 1983. World Textile Trade and Production Trends, Special Report No.152, Economist Intelligence Unit, London.

Caves, Richard, 1985. "Industrial Policy and Trade Policy: A Framework," Tokyo: JERC, mimeo.

Chenery, Hollis and Syrquin, Moises, 1975. Patterns of Development, London: Oxford University Press.

Chenery, Hollis, Sherman Robinson and Moshe Syrquin. Industrialization and Growth: Comparative Study, forthcoming.

Chenery, Hollis and Lance Taylor, November 1968. "Development Patterns Among Countries and Over Time," Review of Economics and Statistics.

Choi, In bom, 1986. "Effects of International Trade on Domestic Market, Structure and Performance," Ph.D. dissertation, Georgetown University.

Chou, Tein-chen, 1985. "Industrial Organization in the Process of Economic Development: Case of Taiwan," mimeo.

Clarke, Davies and Waterson, June 1984. "The Profitability-Concentration Relation: Market Power as Efficiency," The Journal of Industrial Economics.

Corbo, Vittorio, 1985. "Reform and Macroeconomic Adjustment in Chile During 1974-84," World Development.

Corbo, Vittorio and Jaime de Melo, 1985. "Scrambling for Survival: How Firms Adjusted to the Recent Reforms in Argentina, Chile and Uruguay", World Bank Staff Working Paper, No. 764.

Corbo, V., Jaime de Melo and James Tybout, 1986. "What Went Wrong in Southern Cone?" Economic Development and Cultural Change.

Diebold, William, 1980. Industrial Policy as an International Issue, New York: McGraw Hill.

Dornbusch, Rudiger, March 1983. "Remarks on the Southern Cone," IMF Staff Paper.

Edwards, Sebastian, 1983. "The Order of Liberalization of the Current and Capital Accounts of the Balance of Payments," World Bank Working Paper, CPD.

Edwards, Sebastian, 1984. The Order of Liberalization of the External Sector in Developing Countries, Princeton Essays in International Finance, No. 156.

Edwards, Sebastian, January 1985. "Stabilization with Liberalization: An Evaluation of Ten Years of Chile's Experiment with Free Market Policies," Economic Development and Cultural Change.

Edwards, Sebastian and Sweder Van Wijnbergen, February 1986. "The Welfare Effects of Trade and Capital Account Liberalization," International Economic Review.

Edwards, S., 1986, The Korean Macroeconomy: Recent Developments and Future Prespectives, Consultant Report, World Bank.

Engardio, Pete, 1984. "Heavy Industries -- The Winners and Losers," Business Korea, July 1984.

Federation of Korean Banks, The Analysis of the Efficiency of Banking Industry and Suggestions for Higher Efficiency in Korea, 1985.

Frank, Charles, Kwang Suk Kim and Larry Westphal, 1975. Foreign Trade Regimes and Economic Development: South Korea, New York: Columbia University Press.

Freeston, W. Denney and Arpan, Jeffrey S., 1983, The Competitive Status of the U.S. Fibers, Textiles and Apparel Complex: A Study of the Influences of Technology in Determining International Industrial Competitive Advantage, (National Academy Press: Washington, D.C.).

Frenkel, Jacob, 1982. "The Order of Economic Liberalization: Discussion," in Bonner and Meltzer, eds., Economic Poliocy in a World of Change, North Holland.

Frenkel, Jacob, et.al., 1983. "Panel Discussion on Southern Cone," IMF Staff Papers.

Fry, Maxwell, forthcoming. Domestic Resource Mobilization and Allocation through the Financial Sector.

Fry, M., 1982, "Models of Financially Repressed Developing Economies", World Development, No. 10.

GATT, 1984, Textile and Clothing in the World Economy, Background Study for GATT Secretaria, Geneva.

Geber, Anthony, 1985. "The Korea Automobile Industry," Korea's Economy.

Harberger, Arnold, April 1985. "Reflections on the Chilean Economy," Economic Development and Cultural Change.

Hasan, Parvez, 1976. Korea: Problems and Issues in a Rapidly Growing Economy, Baltimore: Johns Hopkins University Press.

Hasan, Parvez and Rao, D.C., 1979. Korea, Policy Issues for Long-Term Development, Baltimore: Johns Hopkins University Press.

Heller, Walter, August 1976. "Change of Factor Endowment and Comparative Advantages: The Case of Japan, 1956-1968," Review of Economics and Statistics.

Hong, Wontack, 1979. Trade, Distortions, and Employment Growth in Korea; Seoul: KOI Press.

Hong, Wontack, 1985. "Import Restriction and Import Liberalization in Export-Oriented Developing Economy: in Light of Korean Experience," Seoul: Seoul University, mimeo.

Johnson, Chalmers, 1982. MITI and the Japanese Miracle: The Growth of Industrial Policy, 1927-1975, Stanford University Press.

Jones, Leroy and Il Sakong, 1980. Government, Business, and Entrepreneurship in Economic Development: The Korean Case, Studies in the Modernization of the Republic of Korea, 1945-1975, Cambridge, Mass., Harvard University Press.

Kalantzopoulos, O. (1986), The Costs of Voluntary Export Restraints for Selected Industries in the US and EEC, Mimeo, World Bank, Washington, DC.

KDI, Long-Term Development Plan for the Year 2000, September 1985.

KDI, Long-Range Prospects for Manpowr Supply and Demand and Policy Tasks, December 1985.

KIET, Projections of Manufacturing Structure and Development Strategy Toward 2000, October 1985.

Kim, Jae Won, 1985. "The Rate of TFP Change in Small and Medium Industries and Economic Development: The Case of Korea's Manufacturing," Korea Development Institute.

Kim, Kwang Suk, 1985. "The Timing and Sequencing of a Trade Liberalization Policy -- The Korean Case," Washington: World Bank, mimeo.

Kim, Kwang Suk and Larry Westphal, 1977. "Industrial Policy and Development in Korea," World Bank Staff Working Paper, No. 263, Washington, D.C., The World Bank.

Kim, Kwang Suk, and Michael Roemer, 1979. Growth and Structural Transformation, Cambridge: Harvard University Press.

Kim, Kwang Suk and Sung Duck Hong, 1982. "Long-Term Variation of Nominal and Effective Rates of Portection," Korea Development Institute.

Kim, Kwang Suk and Joon Kyung Park, 1985. Sources of Economic Growth in Korea: 1963-1982, Korea Development Institute.

Kim, Y.B., 1984. Import and Export Functions in Korea, International Monetary Fund, DM/84/77.

Kojima, Kiyoshi, 1978. Japanese Direct Foreign Investment, Tuttle.

Korean Traders Association, 1982. Foreign Trade Procedures in Korea, Seoul: KOTRA.

Krueger, Anne, 1977. Growth, Distortions, and Patterns of Trade among Many Countries, Princeton Studies in International Finance, no. 40.

Krueger, Anne, 1978. Foreign Trade Regimes and Economic Development: Liberalization Attempts and Consequences, Cambridge, Mass.

Krueger, Anne, 1979. The Development Role of the Foreign Sector and Aid, Cambridge: Harvard University Press.

Krueger, Anne, 1985. "Problems of Liberalization" in Harberger, ed. World Economic Growth.

Krueger, Anne, et al., 1985. Export-oriented Development Strategies: The Success of Five Newly Industrializing Countries, Westview Press.

Kuznets, Paul, 1977. Economic Growth and Structure in the Republic of Korea, New Haven: Yale University Press.

Kwack, Sung Yung, 1985. "Policy Analysis with a Macroeconomic Model of the Korean Economy," Consultant Report, World Bank.

Kwack, Sung Yung and Michael Mered, 1980. "A Model of Economic Policy Effects and External Influences on the Korean Economy," SRI/Wharton EPA World Economic Program Discussion Paper No. 9.

Kwack, Tae Won, 1984. "Industrial Restructuring Experience and Policies in Korea in the 1970s", KDI Working Paper No. 84-08, Seoul: Korea Development Institute.

Lee, Byung-Jong, July 1985. "Special Treatment for the Shipbuilding Industry?" Business Korea.

Lee, Kyu Uck, 1985. "Conglomerates and Business Concentration: The Korean Case," Korea Development Institute.

Lee, Kyu Uck, 1977. "Market Structure and Monopoly Regulation in Manufacturing Industry," Korea Development Institute.

Lee, Kyu Uck and Jae Hyung Lee, 1985. "Business Integration and Concentration of Economic Power," Korea Development Review.

Lee, Sung Sup and Sung Yoon Kang, 1985", Analysis of Export Support System in the Context of Stabilization Regime", Korea Development Institute Review, Fall.

Lee, Young Ki, 1985. "Conglomeration and Business Concentration -- The Korean Case," Seoul: Korea Development Institute, mimeo.

Lee, Young Sun, forthcoming. "Changing Export Patterns in Korea, Taiwan and Japan," Weltwzrstschaffliches Archiv.

Lee, Won-young, 1984. "Science and Technology Policy in Korea," Seoul: Korea Development Institute, mimeo.

Little, I., T. Scitovsky and M. Scott, 1970. Industry and Trade in Some Developing Countries, Oxford University Press.

Magaziner, Ira and Thomas Hout, 1980. Japanese Industrial Policy, Berkeley, University of California, Institute of International Studies.

McKinnon, Ronald, 1973. Money and Capital in Economic Development, Washington D.C., The Brookings Institute.

McKinnon, Ronald, 1982. "The Order of Economic Liberalization: Lessons from Chile and Argentina," in Brunner and Meltzer, eds., Economic Policy in a World of Change, North Holland.

MOST, Science and Technology Policy in Korea, 1985.

Nam, Chong Hyun, 1971. "Trade, Industrial Policies, and the Structure of Protection in Korea," in Trade and Growth of the Advanced Developing Countries in the Pacific Basin, KDI Press: Seoul.

Nam, Chong Hyun, 1985. "Trade Policy and Economic Development in Korea," Discussion Paper No. 9, Korea University, mimeo.

Nishimizu, Mieko and Sherman Robinson, 1983. "Trade Policies and Productivity Change in Semi-Industrialized Countries," Journal of Development Economics.

Noland, Marcus, 1985. "The Changing Pattern of Korean Comparative Advantage, 1965-1980," Institute for International Economics, mimeo.

Noland, Marcus, 1985. The Changing Pattern of Japanese Comparative Advantage", Johns Hopkins University, mimeo.

OECD, 1983, Textile and Clothing Industries: Structural Problems and Policies in OECD Countries, (Paris).

Okuno, Masahiro and Kotaro Suzumura, 1985. "Economic Analysis of Industrial Policy: A Conceptual Framework through the Japanese Experience," Tokyo, Tokyo University, mimeo.

Pack, Howard and Larry Westphal, 1985. "Industrial Strategy and Technological Change: Theory versus Reality", mimeo.

Park, Yung Chul, 1985. "Economic Stabilization and Liberalization in Korea, 1980-1984," Korea University, mimeo.

Park, Yung Chul, 1985. "Korea's Experience with External Debt Management," Seoul: Korea University, mimeo.

Peck, Merton, Richard Levin and Arika Goto, 1985. "Picking Losers: Public Policy Toward Declining Industries in Japan," mimeo.

Petri, Peter, 1983. Synopsis of Japanese Industrial Policy, World Bank, mimeo.

Rhee, Yung Whee, 1984. "Instruments for Export Policy Administration: Lessons from the East Asian Experience," Washington: World Bank, mimeo.

Rhee, Yung Whee, Bruce Ross-Larsen and Gary Pursell, 1984. Korea's Competitive Edge: Managing the Entry into World Markets, Baltimore: Johns Hopkins University Press.

Shaw, Edward, 1973. Financial Deepening in Economic Development, Oxford: Oxford University Press.

Senoo A., ed., 1983, Industrial Concentration in Contemporary Japan: 1971-80.

Shepherd, G., 1981, Textile-Industry Adjustment in Developed Countries, Thames Essay No. 30, (London: Trade Policy Research Centre).

Shujiro Urata, forthcoming. "Sources of Economic Growth and Structural Changes in China: 1956-81.," Journal of Comparative Economics.

Schultze, Charles 1983. "Industrial Policy: A Dissent," The Brookings Review, Vol. 2 (Fall).

Syrquin, Moshe and Shujiro Urata, 1985. "Sources of Changes in Factor Intensity of Trade," Washington: World Bank, mimeo.

Tho, T.V., 1983, Industrial Policy and the Textile Industry: The Japanese Experience, Journal of Contemporary Business, Vol. 11, No. 1, pp. 113-128.

de la Torre, F., 1984, Clothing Industry Adjustment in Developed Countries, Thames Essay, (London: Trade Policy Research Centre).

de la Torre, Jose, et al., 1978, Corporate Responses to Import Competition in the U.S. Apparel Industry, (Atlanta: College of Business Administration, Georgia University).

United States International Trade Commission, 1983, The Effect of Changes in the Value of the U.S. Dollar on Trade in Selected Commodities, (Washington, D.C.), September.

United States International Trade Commission, 1985. Foreign Industrial Targeting and its Effects on U.S. Industries: Phase III: Brazil, Canada, The Republic of Korea, Mexico, and Taiwan, USITC Publication 1632, Washington: USITC.

Vaeth, Werner, 1985. "The State and Problems of Declining Industries -- The Case of West German Steel Industry," Berlin: Free University of Berlin, mimeo.

Van Wijnbergen, Sweder, 1983. "Stagflationary Effects of Monetary Stabilization Policies: A Quantitative Analysis of South Korea," Journal of Development Economics.

Wade, Robert, forthcoming. "The Role of Government in Overcoming Market Failure: Taiwan, Korea and Japan," in H. Hughes et al., eds., Explaining the Success of East Asian Industrialization, Cambridge University Press.

Westphal, Larry, 1978. "The Republic of Korea's Experience with Export-Led Industrial Development," World Development, Vol. 6, No. 3.

Westphal, Larry, 1982. "Fostering Technological Mastery by Means of Selective Infant Industry Protection," in M. Syrquin and S. Teitel (eds.) Trade Stability, Technology and Equity in Latin America, New York: Academic.

Westphal, Larry, forthcoming. "Book Review on Sources of Economic Growth in Korea: 1973-1982" by K.S. Kim and J.K. Park, Journal of Economic Literature.

Westphal, Larry and Kwang Suk Kim, 1982. "Korea," in Balassa et al., Development Strategies in Semi-Industrial Countries.

Westphal, Larry, Linsu Kim and Carl Dahlman, 1984. "Reflections on Korea's Acquisition of Technological Capability," World Bank Discussion Paper. No. DRD77.

Wolf, M., Glismann, H.H., Pelzman, J. and Spinanger, D., 1984, Costs of Protecting Jobs in Textiles and Clothing. Thames Essay, (London: Trade Policy Research Centre).

Woronoff, Jon, 1983. Korea's Economy: Man-Made Miracle, Seoul: Si-sa-yong-o-sa Publishers.

Yamazawa, Ippei, 1980, Increasing Exports and Structural Adjustment of the Japanese Textile Industry. The Developing Economies, Vol. 18, No. 4.

Yamazawa, I. and J. Nohara, ed., 1985, Foreign Trade and Industrial Adjustment in Asia-Pacific Countries, Institute of Development Economics.

Yoon, Chang Ho, and Kung Uck Lee, Industrial Organization, 1985.

Young, Soogil, 1984. "Trade Policy Reform in Korea: Background and Prospect," Seoul: Korea Development Institute, mimeo.

Young, Soogil, 1985. "Role of Trade Policy in Korea's Economic Development and Problems of Import Liberalization," Korea Development Institute, mimeo.

Young, Soogil and Jung Ho Yoo, 1982. "The Basic Role of Industrial Policy and the Reform Proposal for the Protection Regime in Korea," Seoul: Korea Development Institute, mimeo.

Yun, Choon-Sik, 1985. "High Technology Industries in Korea," Korea Exchange Bank Monthly Review.

Yusuf, Shahid and Kyle Peters, 1984. "Saving Behavior and Its Implications for Domestic Resource Mobilization: The Case of Republic of Korea." World Bank Staff Working Paper, No. 628.

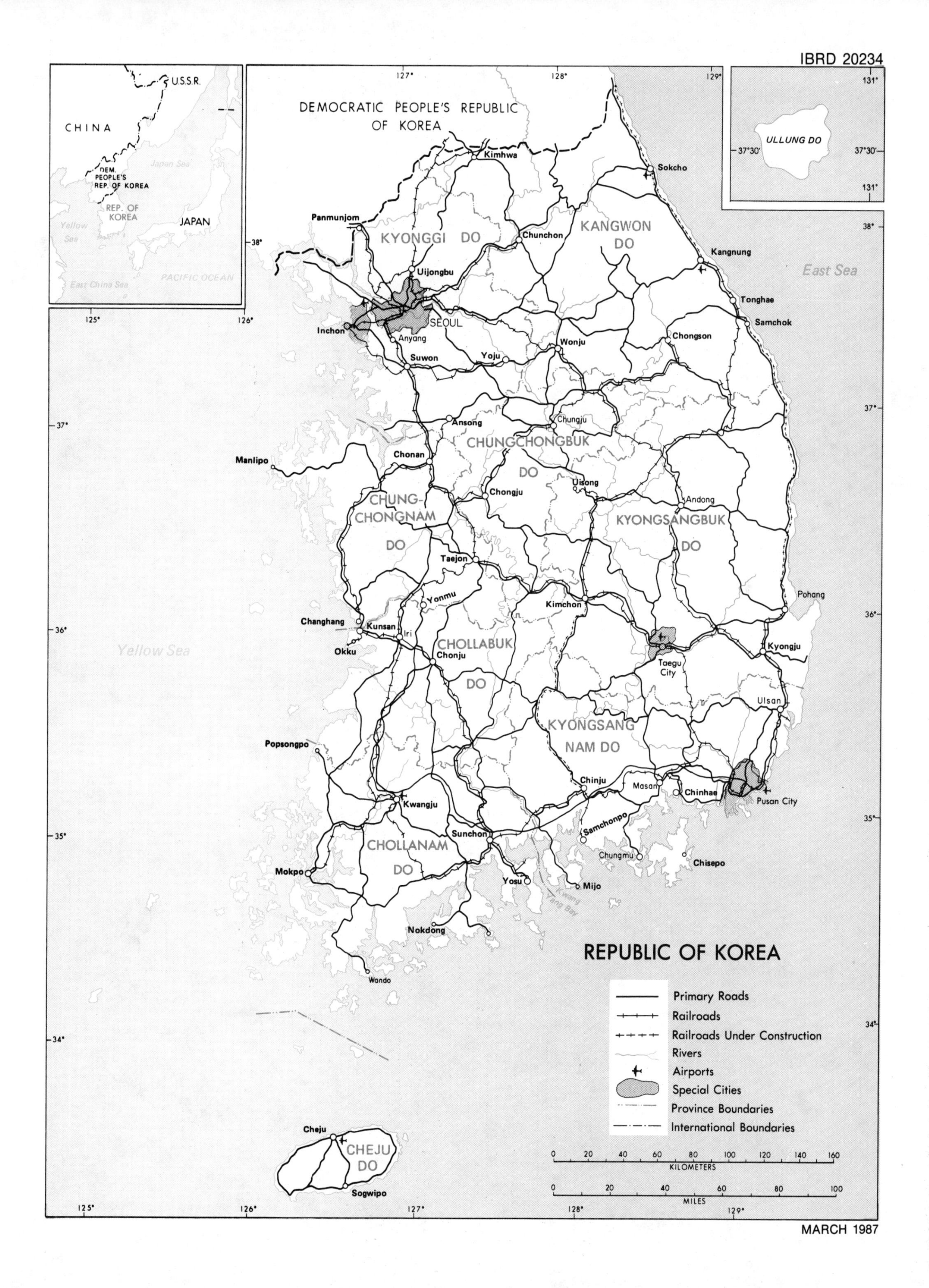
IBRD 20234
U.S.S.R.
CHINA
DEM. PEOPLE'S REP. OF KOREA
Japan Sea
REP. OF KOREA
JAPAN
Yellow Sea
PACIFIC OCEAN
East China Sea
DEMOCRATIC PEOPLE'S REPUBLIC OF KOREA
ULLUNG DO
Kimhwa
Sokcho
Panmunjom
KYONGGI DO
Chunchon
KANGWON DO
Kangnung
East Sea
Uijongbu
Tonghae
SEOUL
Inchon
Anyang
Samchok
Chongson
Wonju
Suwon
Yoju
Chungju
Ansong
CHUNGCHONGBUK DO
Manlipo
Chonan
Uisong
Chongju
Andong
CHUNG-CHONGNAM DO
KYONGSANGBUK DO
Taejon
Pohang
Yonmu
Kimchon
Changhang
Kunsan
Iri
Okku
Yellow Sea
CHOLLABUK DO
Chonju
Kyongju
Taegu City
Ulsan
KYONGSANG NAM DO
Popsongpo
Chinju
Masan
Chinhae
Pusan City
Kwangju
Samchonpo
Sunchon
CHOLLANAM DO
Chungmu
Chisepo
Mokpo
Yosu
Mijo
Kwang Yang Bay
Nokdong
Wando
REPUBLIC OF KOREA
Primary Roads
Railroads
Railroads Under Construction
Rivers
Airports
Special Cities
Province Boundaries
International Boundaries
Cheju
CHEJU DO
Sogwipo
KILOMETERS
MILES
MARCH 1987